Analog and Digital
Electronics
for Scientists

ANALOG AND DIGITAL
ELECTRONICS
FOR SCIENTISTS

Basil H. Vassos
The University of Puerto Rico

Galen W. Ewing
Seton Hall University

Second Edition

A WILEY-INTERSCIENCE PUBLICATION
JOHN WILEY & SONS New York ● Chichester ● Brisbane ● Toronto

Copyright © 1980 by John Wiley & Sons, Inc.

All rights reserved. Published simultaneously in Canada.

Reproduction or translation of any part of this work
beyond that permitted by Sections 107 or 108 of the
1976 United States Copyright Act without the permission
of the copyright owner is unlawful. Requests for
permission or further information should be addressed to
the Permissions Department, John Wiley & Sons, Inc.

Library of Congress Cataloging in Publication Data:

Vassos, Basil H.
  Analog and digital electronics for scientists.

  "A Wiley-Interscience publication."
  Bibliography: p.
  Includes index.
  1. Electronic measurements. 2. Electronic
instruments. I. Ewing, Galen Wood, 1914–   joint
author. II. Title.

TK7878.V38   1979   621.381'028   79-16700
ISBN 0-471-04345-1

Printed in the United States of America

10 9 8 7 6 5 4 3 2 1

PREFACE

The science of electronics has changed extensively during the last few years, particularly in relation to integrated circuits. The development of integrated circuitry has made possible the simultaneous formation of large numbers of interconnected transistors and other components on a single semiconductor chip, at a moderate cost. As a result we are now faced with the paradoxical situation that a one-transistor circuit built with individual components and a 25-transistor integrated circuit cost just about the same. This tends to favor the availability of groups of components that form building blocks (e.g., amplifiers, counters, oscillators, and regulators).

A consequence of this unitary construction is that such devices are very simple to use, normally requiring only power-supply connections, inputs, and outputs; the user can understand and even design circuits to perform complicated functions with very little prior knowledge of electronics.

There are many good books on electronics for the scientist. They seem to be principally of two types. Either they are scaled-down versions of textbooks for engineers, heavily design-oriented, with emphasis on mathematical treatment, or else they are limited to a single area within the field. A selection of useful books is given in Appendix VIII. In contrast, the purpose of the present book is to meet the need of the scientist for a treatment emphasizing the *use* rather than the design of integrated circuits.

In our first edition, we avoided discussing specific models of integrated circuits, feeling that the field was still too fluid. Now, however, it is appropriate to treat selected examples in some detail. We have endeavored to choose for such emphasis devices that are widely disseminated (such as the model 741 operational amplifier) or that have some particular properties that we wish to stress. We should point out that mention of the products of a particular manufacturer does not mean that we consider competitive products to be inferior, but merely that we are more familiar with those described.

This textbook can be used in university classes at both the graduate and

NOV 25 1981

undergraduate levels, since the material is likely to be as new for one group as for the other. In terms of background, one year of calculus and one of general physics are assumed. Some familiarity with differential equations and complex algebra would occasionally be useful, though by no means essential. Otherwise, the text is self-contained; it is designed to be taught in one semester of a lecture course meeting three hours a week, or one full year at two hours a week. With this schedule, the instructor should be able to cover the entire book, or for a simpler treatment, the first nine chapters.

Chapter 12, containing background mathematics, can be intermeshed with various appropriate chapters; alternatively, it can be treated at the end, or be left out entirely.

The experimental portion of the book has been expanded in the second edition. The experiments are graduated in terms of difficulty and the need for original design work on the part of the student. The first few are intended to reinforce the student's understanding of the properties of basic circuit components. Many of the later experiments are based on circuits published in the "Designer's Casebook" section of the magazine *Electronics*.

Subsidiary concepts, not essential to the text, but nevertheless important in their own right, are introduced by way of some of the problems. This is facilitated by the inclusion of detailed answers to half of the problems.

Basil H. Vassos

Galen W. Ewing

Rio Piedras, Puerto Rico
South Orange, New Jersey

CONTENTS

Analog and Digital
Electronics
for Scientists

I

INTRODUCTION

Increasingly since the 1930's laboratory instruments have been designed with extensive use of electronic circuits in both control and data processing functions. Hence, to understand the tools with which they work, scientists and engineers must have at least a rudimentary acquaintance with the principles of electronics. Furthermore, particularly in research, one frequently finds the need of a device to meet specific requirements for which there is no instrument at hand. A working knowledge of basic electronic circuits and components is essential to design a special-purpose instrument.

The development of integrated circuits in the last few decades has simplified greatly both the comprehension of existing electronic circuitry and the design of new circuits. It is for this level of understanding that the present text is intended.

INTEGRATED MODULAR CIRCUITS

The basic simplicity of modern electronics is due almost entirely to the universal application of modularization. In earlier days, to design a circuit, such as a hi-fi amplifier, one had to start with a variety of transistors and painstakingly calculate from formulas (or perhaps determine by trial and error) the required values of associated components. Nowadays, by contrast, a large variety of amplifiers in single, in-

1

tegrated packages, are already at hand at moderate cost.
It is still necessary to utilize additional components
along with these modular units, and their values are
calculated from formulas, but the most difficult parts
of the design have been done by the manufacturer, leaving only the comparatively simple interconnections to
the user.

     As an example, consider a laboratory instrument
such as a pH meter. A model prevalent in the 1960s
contained perhaps 100 discrete components including
several vacuum tubes. A comparable instrument in the
early 1970s used perhaps 10 transistors in place of the
tubes. On the other hand, a present-day pH meter of
the same performance level might have two integrated
(modular) circuit units and only a few additional components, making its electronics much easier to understand.

     The integrated circuits in common use at present
include a wide variety of functional types, in small
plastic or metal housings with protruding wires or
prongs that can be inserted into mating sockets. In
terms of applications, one can distinguish two basic
types, analog and digital. The distinction between
these is important enough to require some clarification
at this point.

     An analog voltage or current (*analog signal*) is a
quantity that can assume any value over a wide range
and is continuously variable. Analog electronic components, then, include amplifiers that change the magnitude of signals, and also functional modifiers that
can, for example, integrate a signal over a time period,
or differentiate it, or take its logarithm.

     Digital signals, by contrast, are inherently dis-
continuous, an example being a series of electrical
pulses, such as the Morse code. Digital signals form
the base of all computer operations. Special hybrid
modules (integrated circuits—ICs) can also be classified according to the magnitude of the circuitry involved. They vary from single-function types, equivalent to a few tens of discrete transistors, through
medium-scale integration (MSI), where up to several
hundred transistors are involved, to large-scale integration (LSI), which can extend to many thousands, all
fabricated at the same time, with all their interconnections, on a single tiny chip of silicon. A typical
IC is depicted in Figure 1-1. The arithmetic and

memory ICs used in pocket calculators are examples of LSI.

Figure 1-1. A typical integrated circuit, shown first as a complete unit, then with its cover removed to expose the actual electronic circuitry. The background is a portion of the artwork for the same device. (Photo courtesy of Analog Devices)

The availability of all these modules marks the disappearance of the mental hurdle previously associated with electronic design in the minds of many scientists. One can now concentrate on designing extensive systems without becoming bogged down in details. This is not to say that no attention need be given to details (otherwise this book would not be necessary), but rather to point out its new emphasis.

Another welcome consequence of integration is the dramatic reduction in the cost of electronic equipment,

whether manufactured or home-made. The price history
of pocket calculators is a well-known example. At the
same time, integrated electronics tends to be much
more reliable than its predecessors.

ELECTRONIC INSTRUMENTS

In this book we are primarily interested in those
aspects of electronics that apply to laboratory instruments. Typically such an instrument is designed to
accept an electrical signal coming from some device
(called a *transducer*) that senses its environment. It
then processes or manipulates this signal in some way
to optimize its yield of information; this can be
accomplished by changing the form of the signal to
render it more intelligible, or by removing unwanted
interference (noise). The resulting modified signal
drives a "read-out" device such as a recorder.

Instruments that fit into this broad pattern are
ubiquitous, and hence the concepts and methods discussed in this book are equally useful to scientists
in nearly any discipline where physical measurements
are made.

II

SIGNALS

Since nearly all aspects of instrumentation are concerned with the generation, processing, and measurement of signals, it is important that the reader thoroughly understands what is meant by the term "signal." This chapter discusses those fundamental properties of signals that are pertinent to instrumentation.

For our purposes, a signal is best defined as a flow of energy that carries information. The *carrier* can be an electric current, a radio wave, or a light beam; it could even be a piece of paper. A steady current or a steady light beam itself carries no information, any more than does a blank sheet of paper. Just as the paper must be written on to provide the information content, so must the current or beam of light be varied in some way from its normal condition if it is to carry information. This process of endowing the carrier with information is called *modulation*.

Typically the modulated carrier is manipulated in various ways, perhaps amplified and made to flow over long distances, before the information is to be retrieved. The final step must be the inverse of modulation, extracting the information from the carrier. This is known as *demodulation*. Figure 2-1 shows these relations.

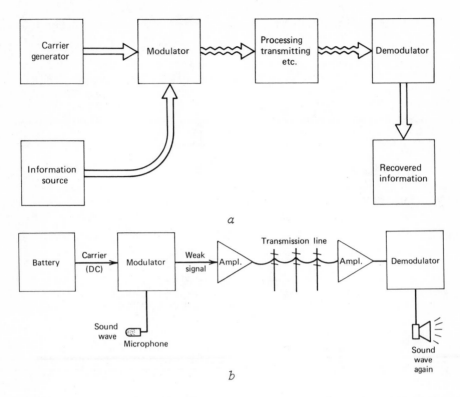

Figure 2-1. (*a*) A generalized signal system. (*b*) A specific embodiment, the transmission of voice over a telephone system.

CARRIERS

For most of the applications considered in this book, the carrier of information will be electrical. There are, however, important applications of other types of carriers. A light beam is one of these. Research is presently underway in the transmission of voice signals by modulation of a light beam to be propagated through long glass fibers; it is expected that this will greatly increase the efficiency of telephone service.

Another application involving carrying information by light is the *opto-isolator*, a miniature device consisting of a tiny light source and a photosensor

encapsulated together. A modulated electric current
entering the device will pass on its modulation to the
light, and the photosensor will convert it back to the
electrical domain as output. Since there is no direct
connection between input and output, they are electrically isolated from each other, with, nevertheless,
efficient transfer of information.

Many signal concepts were originated in connection
with radio communication. The carrier in this case is
a high-frequency electromagnetic wave. The modulator
is an electronic circuit in the transmitter that mod-
ifies the carrier wave in response to signals from a
microphone, while the demodulator is a circuit in the
receiver that retrieves the information and converts
it back to audible form.

ELECTRICAL QUANTITIES

The electrical quantities of primary concern to us
are listed in Table 2-1, together with the name of the
basic unit and commonly used symbols. A distinction
is sometimes made between instantaneous voltages and
currents (e, i) and the steady-state quantities (E, I).
In this book the symbol V is mostly used for power
supply voltages.

A voltage or current that is constant is referred
to as DC; if it is periodically time dependent, it is
called AC. A nonrepetitive, time-dependent signal is
described as a *transient*.

The basic variety of AC is the sine-wave. In contrast to DC voltages and currents, which are completely
defined by two quantities, their sign and magnitude,
sine waves need *three* quantities for a complete description: amplitude (A), frequency (f), and phase
(ϕ), as shown in Figure 2-2. Since the signal has various amplitudes at different moments, it is convenient
to define an average quantity called the root-mean-
square (RMS) voltage:

$$E_{RMS} = \sqrt{\overline{E^2}} \qquad (2-1)$$

where the bar over the E^2 indicates the time average.
A similar formula is valid for the RMS current. For

TABLE 2-1

Electrical Quantities[a]

Quantity	Unit	Symbol
Potential	Volt (V)	E, e, V
Charge	Coulomb (C)	Q
Current	Ampere (A)	I, i
Power	Watt (W)	P
Resistance	Ohm (Ω)	R
Conductance	Siemens (S)	g
Impedance	Ohm (Ω)	Z
Admittance	Siemens (S)	G
Capacitance	Farad (F)	C
Inductance	Henry (H)	L
Frequency	Hertz (Hz)	f

[a] The usual multipliers (k, m, M, μ, etc.) are used with the unit abbreviations. One is likely to encounter in the literature various other abbreviations for the microfarad (mfd, MFD, uf, etc.) and for the picofarad (μμF, mmf, uuf, etc.). The siemens is often given the older name "mho" with the abbreviation Ω^{-1}.

sine waves the RMS value is $1/\sqrt{2}$ or 71% of the peak value A. The line voltage is conveniently expressed as its RMS value: "115 V RMS" means $115 \times \sqrt{2} = 162V$ amplitude (A), in both positive and negative directions, or 324 V peak-to-peak.

The power in an AC circuit has a pulsed character-istic since it must cross zero whenever the sine wave crosses zero, twice every cycle. The RMS voltage

represents the value of an equivalent DC voltage that
would generate the same heating effect when applied to
a resistor.   Thus a 115-V AC (RMS) and a 115V DC will
provide equal total heat in a kitchen range, even
though the former comes in pulses.

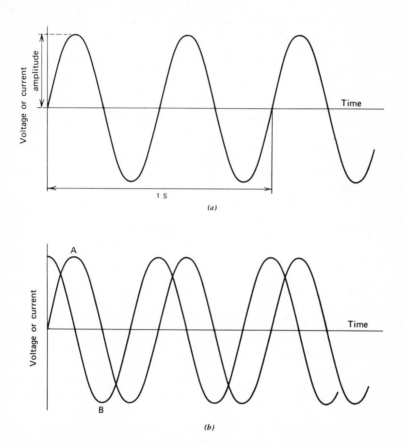

Figure 2-2. (*a*) An example of an AC signal;   frequency = 2 Hz.
The number of oscillations in 1 sec is the frequency.   (*b*) Phase
difference between two signals of equal frequency and amplitude.

        Figure 2-2*b* shows that two signals of the same
frequency and amplitude can have different values at
the same instant.   Signal B reaches any particular
value earlier than does A;   B is said to *lead* and A to
lag in phase.

In general, a sinusoidal wave can be described by
the expression

$$E = A \sin (2\pi f t + \phi) = A \sin (\omega t + \phi) \qquad (2\text{-}2)$$

where t is the time, A the amplitude, f the frequency,
and ϕ the phase difference with respect to some reference (see Figure 2-3). The rate of rotation of the
radius vector $\overline{OM}$ is ω rad/sec ($2\pi f = \omega$). If two signals have the same values of f, but differ in ϕ, they
will appear displaced with respect to each other along
the time axis, as in Figure 2-2b. Usually this displacement is not expressed in time units, but in angles
of rotation of the vector. For example, a difference
of 180° (π radians) means that the two waves are exactly opposite to each other (out of phase). The waves
in Figure 2-2b differ by 90° or $\pi/4$ rad, and thus are
said to be 90° out of phase. Note that curve B is a
cosine wave.

Many AC signals are more complex than a sine wave.
It can be shown that essentially all periodic signals
can be represented as sums of sine and cosine functions
(*Fourier expansion*).

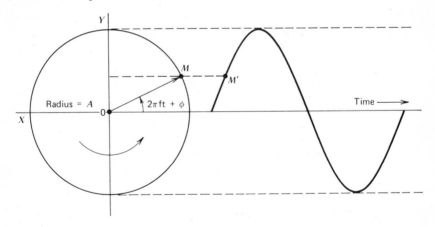

Figure 2-3. The generation of a sine wave from circular motion.
The radius *OM* (= *A*) can be considered to be a vector rotating
about the origin in the counterclockwise direction. The position
of *M*, projected on the *Y* axis, becomes a sine wave when represented as a function of time.

In addition to voltage and current, we are also interested in *power* representing the heat generated in a device defined for DC quantities as the product of current and voltage:

$$P = EI \qquad (2-3)$$

By using Ohm's law, one can also show that

$$P = \frac{E^2}{R} = I^2 R \qquad (2-4)$$

With AC, the phase relationship between E and I also plays a role. If the phase difference is ϕ, then the power is given by

$$P = EI \cos \phi \qquad (2-5)$$

where E and I are RMS quantities. The factor $\cos \phi$, which can vary from zero to one depending on the components involved, is often designated as the *power factor*.

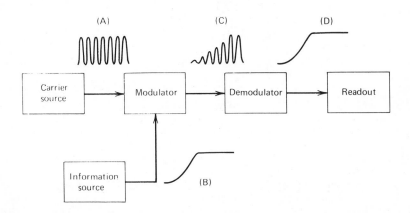

Figure 2-4. Amplitude modulation. The carrier is a relatively high-frequency continuous wave (A). The shape of the information is shown at (B). The modulated wave (C) has the same frequency as (A), but its amplitude envelope is determined by (B). Various signal-processing steps are usually inserted between modulator and demodulator. Note that the wave forms sketched are actually graphs plotting the appropriate voltages against time, a procedure often employed in electronic schematics.

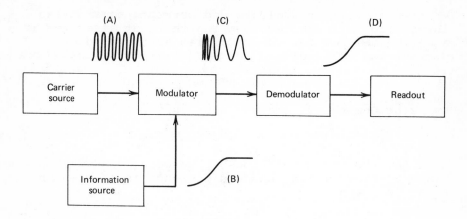

Figure 2-5. Frequency modulation. The block diagram is exactly the same as for Figure 2-4, but the waveform at (C) is different, showing a frequency that changes in accordance with information at (B).

MODULATION

In radio transmission two types of modulation are common: *amplitude modulation* (AM) and *frequency modulation* (FM). In laboratory instrumentation both are used, but AM is by far the more common. In AM the amplitude of the AC carrier is made to follow variations in signal level, as diagrammed in Figure 2-4. In FM the frequency of the carrier rather than its amplitude is varied (Figure 2-5). There are several other less common types of modulation that will not be introduced at this point.

The usefulness of modulation is in improving the freedom from noise, both in measurement systems and in data transmission. Of the two methods discussed, FM exhibits superior noise elimination, but requires more complex equipment than does AM.

AMPLIFICATION AND ATTENUATION

Amplification in electronics, paralleling the everyday use of the word, refers to a process by which a signal is increased in magnitude. The quantity to

be amplified can be a voltage, a current, or a power,
thereby defining respectively a voltage, current, or
power gain (see Figure 2-6). Since the power in an
electric circuit is equal to the product of current and
voltage, it follows that the power gain of an amplifier
is equal to the product of its voltage and current
gains:

$$\frac{P_{out}}{P_{in}} = \frac{E_{out}}{E_{in}} \times \frac{I_{out}}{I_{in}} \qquad (2\text{-}6)$$

Note that it is quite possible for a particular amplifier to have a voltage (or current) gain of unity or
even less, while still having a large power gain.

An amplifier must be an *active device*, meaning
that it must extract power from an external source,
such as a battery, to increase the power in the signal.
No *passive* device (one without access to a power source)
can do this.

An *attenuator* can be considered the inverse of an
amplifier, diminishing the voltage or current supplied
to it. The most familiar example of attenuator is the
volume control of a radio. Both voltage and current
gains (hence also the power gain) are less than one.

Power gains and attenuations are usually expressed in terms of a logarithmic unit, the power *decibel*,
dB_p, defined in terms of input and output levels P_1
and P_2:
$$dB_p = 10 \log \frac{P_2}{P_1} \qquad (2\text{-}7)$$

This applies equally well to gains (where $P_2 > P_1$) and to
losses (or attenuations, where $P_2 > P_1$). For example,
$dB_p = -20$ indicates an attenuation of 100, whereas dB_p
$= +30$ means a power amplification of 1000.

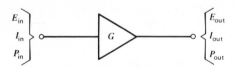

Figure 2-6. General symbol for an amplifier. G is the gain,
equal to the ratio of output to input quantities.

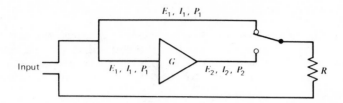

Figure 2-7. Amplifier connections such that the resistor R can be connected either directly to the input source or through an amplifier.

The dB notation can be extended to voltage and current ratios with some modification. To see how this should be done for voltage ratios, recall that the power P dissipated in a resistor of R ohms is given by $P = E^2/R$. Consider now a resistor that can be connected either directly or through an amplifier to a source of power (Figure 2-7). The switch thus permits one to select either input or output power to be applied to the resistor. For this circuit we can write input and output powers as

$$P_1 = \frac{E_1^2}{R} \text{ and } P_2 = \frac{E_2^2}{R} \qquad (2\text{-}8)$$

Then, substituting these relations into Eq. (2-7) gives

$$dB = 10 \log \frac{P_2}{P_1} = 10 \log \frac{E_2^2/R}{E_1^2/R} \qquad (2\text{-}9)$$

This simplfies to:

$$dB = 20 \log \frac{E_2}{E_1} \qquad (2\text{-}10)$$

Hence the voltage gain can be expressed logarithmically by

$$dB_E = 20 \log \frac{E_2}{E_1} \qquad (2\text{-}11)$$

An exactly analogous expression applies to the current gain:

$$dB_I = 20 \log \frac{I_2}{I_1} \qquad (2\text{-}12)$$

Notice that the above definitions predict that

$$dB_p = \frac{dB_E + dB_I}{2} \qquad (2\text{-}13)$$

The gain as calculated above is the so-called *insertion gain*, and not the input-to-output gain of the amplifier itself, which depends upon the input resistance of the amplifier, to be defined later.

The decibel notation has a number of advantages and should be mastered by anyone pursuing the study of electronics. The first advantage is the use of small numbers rather than large, particularly where no great precision is needed. Thus an amplifier with a voltage ratio of 1,000,000 to 1 has a gain of 120 dB. Another advantage is the convenient additivity of gains and losses in decibels. For example, if a 120-dB amplifier were modified by adding an attenuator with a 10-dB insertion loss, the net gain of the combination would be merely 120 - 10 = 110 dB.

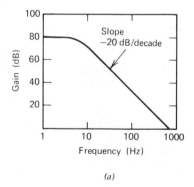
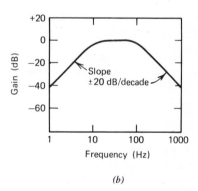

(a)   (b)

Figure 2-8. Examples of frequency-response curves for (a) an amplifier, (b) an attenuator. The frequency-selective networks responsible for the sloping portions of these curves often produce a roll-off of 20 dB per decade, as shown. An advantage of such representations is that in most cases the curves consist of a few intersecting straight lines, either horizontal or at a slope of 20 dB per decade.

Gain or attenuation expressed in decibels is often plotted as a function of frequency, the latter also on a logarithmic scale, as in Figure 2-8. This sometimes is called a *Bode plot*, and is used for displaying the frequency response of any device. The use of logarithmic scales on both axes makes it possible to display a very wide range of frequencies and gain ratios on a single sheet of paper. Such plots are used extensively in many areas of electronics and will be encountered frequently in this book.

TRANSDUCERS

In the majority of instruments, the objective is to measure some property of an external system. The device that converts the information from the external source into electronic signals is the *transducer*. Some transducers are able to sense the desired property by themselves. A thermocouple is a good example; it produces a voltage proportional to the temperature, without the need of any additional source of electrical power. On the other hand, a resistance thermometer requires a current from an outside supply to produce a voltage drop across the temperature-sensitive element, as a resistor cannot produce signals directly. Note the analogy with active and passive devices.

Sometimes the transducer by itself cannot provide all the information desired. In such cases a sudden perturbation (*excitation*) must be used to produce a measurable response. Optically excited fluorescence falls in this category; a specimen must be irradiated with ultraviolet for the fluorescent emission to occur. Similar situations are frequently encountered in biomedical experimentation, where the responses to stimuli are observed. Electronics is likely to be involved in both the excitation and measurement steps.

TRANSFER CHARACTERISTICS

The signal from a transducer is normally fed into an electronic device that generates some function of its input. As an example, an amplifier may be fed a signal of 1 mV to produce an output of 10 V. A unit to fill this requirement can be implemented in numerous ways with the same end result. Consequently the user can simply think of it as a "black box" character-ized by a gain of 10,000, without the need to consider what is inside. This concept is widely used to sim-plify electronic thinking, since the action of any "black box" that accepts a signal and puts out a re-lated one can be completely described by a relation of the form

$$S_{out} = f(S_{in})$$ (2-14)

where S_{out} and S_{in} represent output and input signals, respectively, and f is some function, called the *trans-fer coefficient*.* The general symbol S is to be re-placed by E or I according to whether the signal is a voltage or a current.

The transfer coefficient is often simply a con-stant multiplier, and the circuit is said to be *linear*. In other cases $f(S_{in})$ may be a logarithmic, quadratic, or other mathematical function.

In practice, f remains constant only over a finite range of the variable S. This may be made clear by considering an example. Suppose we have an amplifier that is designed to be linear with a gain of 10,000, hence following the relation

$$E_{out} = 10^4 \cdot E_{in}$$ (2-15)

If this is a modern solid-state amplifier, E_{out} will probably be limited physically to the range -10 to +10V. It follows from the equation that the input must be restricted to the range 0 to $10/10^4$ V or 0 to 1 mV. If a voltage greater than this is applied to the input, the output will not be able to respond linearly, but will level off (*saturate*) at 10V. At the other ex-treme, as the input approaches zero, linearity is again lost, but for a different reason. All electrical sys-tems contain random fluctuations called *noise*. The noise in well-designed circuits is generally small compared to bona fide signals, but as the signal is

*Note that the term "transfer function" has another meaning, and is not to be used in the present context.

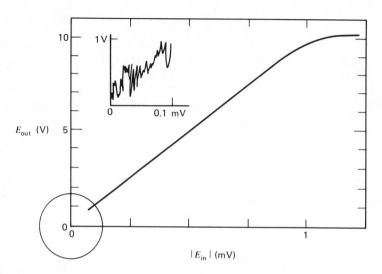

Figure 2-9.   Measured transfer plot for a linear amplifier as described in the text.   The noise-limited region is shown in the inset.

reduced, the noise becomes significant and eventually limits the sensitivity. Figure 2-9 shows the *transfer plot* (E_{out} as a function of E_{in}) for an amplifier, illustrating the nonlinearities at both ends.

The extent of the linear region is known as the *dynamic range* of the amplifier, in this case extending from approximately 0.1 mV to 1 mV, referred to the input. If this were a logarithmic amplifier, the term dynamic range would describe the region over which the log function is accurately followed. Occasions will arise where the dynamic range is limited by other phenomena than saturation and noise.

A major objective in all sorts of circuits that handle signals is to increase the signal-to-noise ratio, often denoted by the symbol S/N. If the output of a transducer shows an S/N of less than 2, the signal is obscured by the noise to the point of being almost useless. The succeeding electronics, if properly designed, could amplify the signal more than the noise, thus increasing the S/N to a useful level, say 20, but plain amplification effects both S and N equally without changing their ratio. Enhancement of S/N is often difficult to accomplish. Various techniques for doing this will be considered in later chapters.

NOISE

Noise can arise in many different ways, some of which can be eliminated, but others are unavoidable and can merely be minimized. The basic difficulty is that the noise is observable by the same means as the desired signal. This is why the signal-to-noise ratio is given so much attention.

Noise is usually observed at the output of an amplifier, but for a quantitative measure is conveniently referred to the input by means of a parameter called the *noise equivalent power* (NEP). Thus an amplifier with a power gain of 1000 and 10 mW of noise at the output has an NEP of 10 µW.

There are four main categories of noise of importance to us: resistance noise, shot noise, current noise, and environmental noise. The first two are inherent in the electronic systems themselves, the third results from properties of particular components, and environmental noise is produced by external phenomena.

Resistance noise, also called *Johnson noise*, results from the random thermal motion of electrons in resistive components. The voltage that it produces, squared and averaged over a period of time, is given by the simple expression

$$\overline{e_{nR}^2} = 4kTR_s \Delta f \tag{2-16}$$

where k is the Boltzmann constant (1.38×10^{-23} J/K), T is the Kelvin temperature, R_s is the source resistance in which the noise originates, and Δf represents the width of the band of frequencies over which the measurement is made (*bandwidth*). This means that within a particular frequency span—say 100 Hz—the resistance noise will be the same no matter what part of the spectrum is selected. (The term *white noise* is used in this case by analogy with white light, which has a uniform distribution of frequencies.)

For a given resistance value, the square of the voltage is proportional to power; hence we can rewrite Eq. (2-16) as

$$\overline{P_{nR}} = \frac{\overline{e_{nR}^2}}{R_s} = 4kT\Delta f \tag{2-17}$$

which gives the rather surprising result that resistance noise *power* is independent of the resistance.

Resistance noise voltage can be reduced only by
lowering the values of R or T or both (frequently impossible or cumbersome) or operating with a small bandwidth.

Shot noise (Schottky noise) is observed whenever
a current passes through some interface, (for instance
in a transistor). It reflects the fact that electricity is quantized and can only flow in units of single
electrons. Shot noise is usually specified as a random fluctuation i_{nS} of the current I, again squared
and averaged over a time interval:

$$\overline{i_{nS}^2} = 2I\epsilon\Delta f \qquad (2\text{-}18)$$

The corresponding power expression is

$$\overline{P_{nS}} = 2IR_L\epsilon\Delta f \qquad (2\text{-}19)$$

The R_L refers to the load resistance through which the
current $I \pm i_{nS}$ is flowing and ϵ is the electronic
charge, 1.602×10^{-19} C. Note that shot noise is also
white, and is dependent on the bandwidth, Δf.

Current noise, or *flicker noise*, originates by a
somewhat involved mechanism in resistors and other
components that are granular in composition. It is
not white, but has an inverse dependence on the frequency:*

$$\overline{e_{nC}^2} \propto \frac{\Delta f}{f} \qquad (2\text{-}20)$$

$$\overline{P_{nC}} \propto \frac{\Delta f}{R_L f} \qquad (2\text{-}21)$$

* The fact that the expressions of Eqs. (2-20) and (2-21) become
infinite for $f = 0$ need not disturb us, since zero frequency implies an infinite period, not realizable experimentally. DC is
only an approximation to zero frequency.

Current noise is found to some extent in all components,
but is especially marked in transistors, carbon-paste
resistors, and photocells. Metallic film and wire-
wound resistors show the effect to a smaller degree.

The $1/f$ dependence suggests that this type of
noise can be important at very low frequency, including
in the limit zero frequency (DC). This is indeed true,
and means that it is essential to avoid DC measurements
when working with very small signals.

One of the most annoying phenomena encountered
with low-level DC signals is that of *drift*. This is a
continuing change in output caused by some minor variation in the circuit. It is most often the result of
slow temperature changes, but may also represent aging
of components. It can be considered as $1/f$ noise in
the limit as the frequency approaches zero.

Another effect that may seem to resemble drift is
voltage offset. ideally an amplifier should give zero
output for zero input signal. If nevertheless a voltage appears, it is called offset. It differs from
drift in that it is of constant value. The cause of
offset is usually a slight misadjustment in the input
section. It can be compensated by the addition of a
small opposing voltage.

Impulse noise appears in the form of undesirable
impulses. There are several types of impulse noise.
One of these, due to switching transients, is especially noticeable if both analog and digital circuitry
are utilized in close proximity. Digital circuits are
characterized by sudden shifts in voltage from one
level to another, often synchronized over a whole system. At the moment of switching, an impulse is produced that can be sensed by nearby analog components.
This type of interference is best minimized by physical
separation and shielding of analog and digital circuits
and the careful use of ground connections.

Another variety of impulse noise is called "*pop-
corn noise*" because in an audio system it sounds like
popping corn. This originates within the integrated
circuits; specially processed ICs may have little of
this type of noise.

Most other types of noise can be lumped together
under the heading of *environmental noise*. Figure 2-10
suggests qualitatively the various kinds of interferences to be expected in a typical university laboratory.

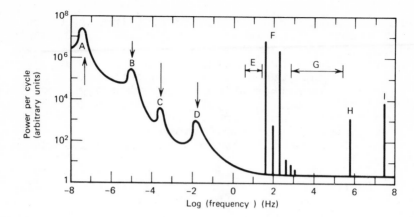

Figure 2-10. The $1/f$ dependence of environmental noise. [From Coor, *J. Chem. Educ.*, 45, A583 (1968), an excellent short treatment.] The noise sources are identified as follows: (A) temperature variations (per year); (B) temperature variations (per day); (C) change of classes (per hour); (D) elevator operation (per minute); (E) fairly noise-free region; (F) power-line frequencies (60, 120, 180, etc., Hz); (G) good quiet region; (H) AM radio interference; (I) television interference.

Observe that below perhaps 10 Hz such noise appears to show an inverse frequency response, similar to current noise. It can be minimized by the same method—conversion to AC.

   A special case that can be particularly troublesome is "pickup" of spurious signals at the frequency of the power line and its first few harmonics. This pickup can occur through a direct conductive path (e.g., leakage through a cracked insulator) or more likely by magnetic or capacitive coupling. It can be controlled by a combination of carefully planned grounding and shielding, which is discussed in a later chapter.

PROBLEMS

2-A. By way of review, write all general relations possible between the quantities in Table 2-1.

2-B. An amplifier, upon receiving 1-mA input current, gives an output of 0.5 V. What is the transfer coefficient? In what units?

2-C. Consider an amplifier fed from a source whose noise can be neglected. Assume the input noise of the amplifier to be of the Johnson type only. For a given bandwidth, the average noise power, $\overline{P_{nR}}$, is 10^{-14}W. Compute the average voltage, $\left[\overline{E_{nR}^2}\right]^{\frac{1}{2}}$, and current noise, $\left[\overline{I_{nR}^2}\right]^{\frac{1}{2}}$, for R_{in} = 100, 1000, and 10,000 Ω.

2-D. Consider an amplifier with a gain of 32 dB, followed by a filter with insertion loss (attenuation) of 12 dB at the frequency of interest, followed by a voltage divider formed from a 100 Ω and a 60 Ω resistor. What is the overall gain if the output is taken across the 60-Ω resistor?

*   *   *

2-1. Prove that the RMS value of a sine wave is $E_o/\sqrt{2}$, by means of Eq. (2-1), noting that the average of any periodic function $f(\theta)$ over the interval of zero to 2π is $\overline{F} = (1/2\pi) \int_0^{2\pi} F d\theta$.

2-2. What is the RMS voltage of the square wave shown in Figure 2-11?

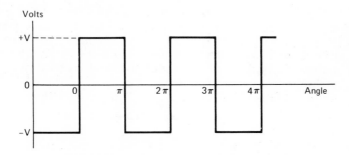

Figure 2-11. See Problem 2-2.

2-3.   In a particular component there is a current $I =$ 0.1 cos (240t + 45°) and a voltage E = 10 cos (240t - 135°) where I is in amperes, E in volts, and t in seconds.

     (a)   What is the frequency in hertz?
     (b)   What is the phase difference between E and I in degrees and in radians?

2-4.   Calculate the peak-to-peak and RMS values of both E and I in Problem 2-3.

2-5.   Compute the power dissipated in a component where I = 0.3 sin (63t + π) and E = 3.0 sin (63t + π/2). What is the power factor?

2-6.   Compute the decibel values for the following ratios of E_{out}/E_{in}:   (a) 0.01, (b) 0.1, (c) 1.0, (d) 100, (e) 2, (f) 3.142, and (g) 90.

2-7.   Consider two circuits in series, one (A) with an attenuation of -20 dB, and the second (B) with -3dB.   Show mathematically that the overall attenuation is -23 dB, the *sum* of the individual attenuations.

2-8.   Suppose that a circuit has an attenuation of -73 dB.   In the decibel table (Appendix III) we can find the ratios corresponding to 60, 10, and 3 dB.   Show that the overall voltage ratio is obtained by taking the *product* of the ratios for the three individual decibel values.

2-9.   A particular resistor is labeled as 100 Ω at 2 W. The wattage rating indicates the maximum power that can be dissipated in the resistor without overheating.   What is the maximum DC voltage that can safely be applied across the resistor?

2-10. An amplifier of gain A_p = 100   generates an output noise of 1 mV into a load of 1000 Ω.   What is its NEP figure?

2-11. An amplifier has a voltage noise referred to the input of 0.1 mV/Hz$^{1/2}$.   Compute the signal-to-noise ratio at the output if an input AC signal of 100 mV is applied and the bandwidth of the amplifier is (a) 100 Hz, (b) 10 Hz, and (c) 1 Hz.

III

PASSIVE COMPONENTS AND CIRCUITS

The heart of any electronic circuit lies in its active components, but no active device can function alone. It must have a selection of passive units to support it. This chapter is intended to provide insight into the properties and modes of operation of these passive devices, so that they can be used intelligently when needed.

The passive components to be considered are resistors, capacitors, inductors, and diodes. The properties of capacitors and inductors are strongly dependent on the frequency of alternating current with which they are used or on the rate of change of direct current. They show no response to unchanging DC. Resistors, on the other hand, are generally equally useful in AC or DC circuits, and it is convenient to discuss them first. The response of diodes is more complex.

RESISTORS

Resistors obey *Ohm's law:* $E = RI$, where R is a proportionality constant called the *resistance*. The resistor as a device is characterized not only by a well-defined value of its resistance,* but also by its

*The word *resistor* refers to the physical object or component; *resistance* refers to the magnitude of the property of a resistor known as its resistance. So we speak of adding resistances in series, or of connecting resistors end-to-end.

power rating,* and temperature coefficient. Commercial
fixed-value resistors are of several types: carbon or
composition resistors, which are the least expensive;
metallic film types, which have less tendency to introduce noise into the system; and wire-wound units, including those capable of dissipating considerable power
as heat (about 5 watts and greater).

Variable resistors are also common. These are particularly important as providing a means for adjustment
of a circuit. Variable resistors usually have three
connections, as symbolized in Figure 3-1. If only two
terminals are needed, it is best practice to wire permanently one end to the variable contact or wiper, as
shown in Figure 3-1b.   This arrangement tends to reduce
the noise produced by the wiper.

Commercial resistors are never perfect. Wire-wound
resistors are more stable, but may have enough inductance to affect their operation at high frequencies.
Carbon composition units have negligible inductance,
but higher temperature coefficients, and are more prone
to introduce noise into the circuit. Metal film resistors are the preferred type for high-quality circuits,
but are more expensive.

Combinations of resistors often occur, and it is
useful to compute the associated voltages and currents.
Two rules for combinations are basic: (1) resistances
in series are additive, and (2) resistances in parallel
follow a reciprocal law. These are illustrated in Fig-

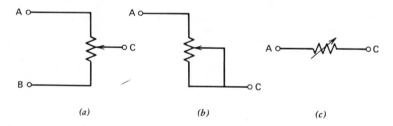

(a)   (b)   (c)

Figure 3-1.   (a) A three-terminal variable resistor, also called
a *potentiometer* or "pot".   (b) The same unit wired for two-
terminal operation, sometimes called a *rheostat*.   (c) An alternative symbol for (b).

*The power rating is the maximum power that a component can dis-
sipate without damage. It is expressed in watts.

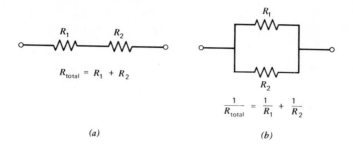

Figure 3-2. Two resistors connected (a) in series, (b) in parallel.

ure 3-2, where it is shown that each combination is equivalent to the single resistor indicated.

As a specific example, consider the network of resistors shown in Figure 3-3a. To simplify this, we first combine R_2 and R_3 by the reciprocal law, to an equivalent, 100 Ω, shown in Figure 3-3b. Then this adds directly with R_4 to give 600 Ω (Figure 3-3c), which is in turn treated as a parallel pair with R_5, giving 300 Ω. The overall resistance then becomes 300 + 150 = 450 Ω.

An important application of these principles is to the *voltage divider* shown in Figure 3-4. In this circuit the input voltage E_{in} is impressed across the two resistors but only a fraction of it appears across R_1 to give the output voltage E_{out}. By Ohm's law, the current I_{in} is given by $E_{in}/(R_1 + R_2)$. This current produces in R_1 a drop of voltage $E_{out} = I_{in}R_1$. By combining these two relations one can obtain the transfer coefficient

$$\frac{E_{out}}{E_{in}} = \frac{R_1}{R_1 + R_2} \qquad (3\text{-}1)$$

This is the voltage divider equation, which we shall repeatedly have occasion to invoke. Two special cases exist, which warrant more detailed discussion.

If the sum of R_1 and R_2 is constant, as happens

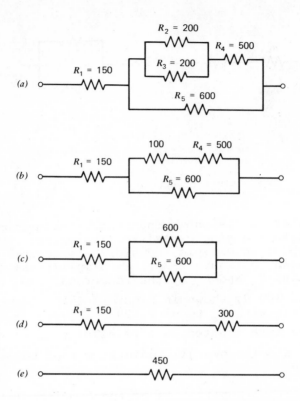

Figure 3-3. A multicomponent resistive network. The original
network (a) is shown resolved stepwise into its equivalent single
resistor (e). Note that the total power dissipation is the same
in all of these circuits, but is unequally distributed between
component resistors.

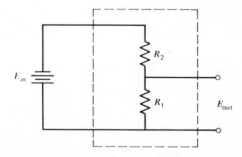

Figure 3-4.   The voltage divider.

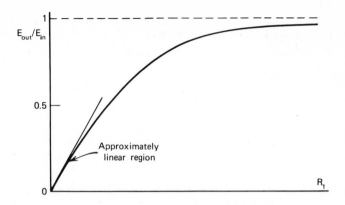

Figure 3-5. The variation in the output from the voltage divider
of Figure 3-4, when R_2 is held constant and R_1 is varied. The
output is nearly linear with changes in R_1 for small values.

when they are the two branches of a potentiometer, then
the output is directly proportional to the value of R_1.
This can be seen by an inspection of Eq. (3-1). Many
potentiometers have R_1 proportional to the rotation of
a shaft. In this case, E_{out} is also proportional to
angular displacement.

Another interesting case arises when only one of
the resistors is variable. Now the total resistance is
also variable, and the transfer coefficient becomes

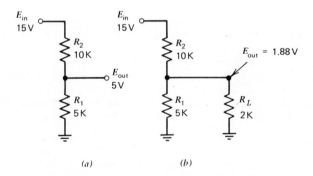

Figure 3-6. A voltage divider, (a) with no load, and (b) with a
load resistor R_L.

somewhat more complicated. From Figure 3-5, it is evident that the relation between E_{out} and R_1 is linear for small values of R_1 as compared to R_2, but deviates for larger values.

LOADING

The output of a voltage divider can be affected by whatever is connected thereto. Consider Figure 3-6a. The output is E_{out} = 15(5000/15000) = 5 V (by Eq. 3-1). This is true only if no current is taken from the output terminal. However, if for example a 2000-Ω load is added (as at b), the output voltage will be greatly diminished. To calculate its value, the resistance of R_1 and R_L in parallel must be substituted for R_1 in Eq. (3-1). This parallel resistance is (5000)(2000)/(5000 + 2000) = 1430 Ω. Then Eq. (3-1) gives us E_{out} = 15(1430/11430) = 1.88 V, obviously much smaller than 5 V. The loading effect becomes negligible only if $R_L >> R_1$. Otherwise it must always be kept in mind. In addition, electronic circuits have input and output resistances, R_{in} and R_{out}, defined by Ohm's law as the ratios of the respective voltages and currents:

$$R_{in} = \frac{E_{in}}{I_{in}} \quad \text{and} \quad R_{out} = \frac{E_{out}}{I_{out}}$$

These resistances could also influence the behavior of voltage dividers, as the following example will show. Consider a typical situation that might occur in the laboratory: the coupling of an instrument such as a gas chromatograph (GC) to a strip-chart recorder that was not designed for it. Suppose the GC has an output range of 0 to 100 mV and an output resistance of 10 Ω; the recorder goes to full scale for 10 mV, and its input resistance is 10^7 Ω. Figure 3-7 shows this situation schematically, and suggests the use of a voltage divider to adapt the GC to the recorder. The internal resistance of the GC must be considered to be in series with R_2 whereas the input resistance of the recorder

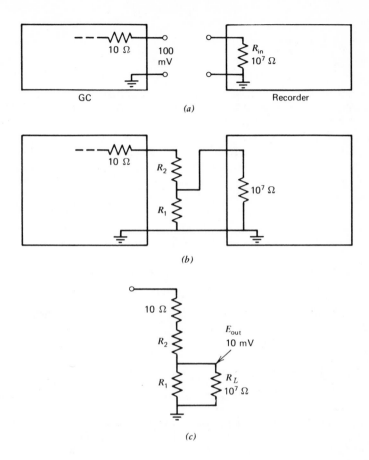

Figure 3-7. A gas chromatograph and recorder to be interfaced so that 100 mV from the GC provides 10 mV to the recorder. (*a*) The problem, (*b*) a voltage divider as a possible coupling device, (*c*) an equivalent circuit.

parallels the lower leg of the divider (R_1). Thus we can draw an equivalent circuit as in Figure 3-7*c*.

The input resistance of the recorder is so large that for practical values of R_1, it exerts no significant loading effect on the voltage divider. Therefore Eq. (3-1) can be applied without correction:

$$\frac{R_1}{R_1 + R_2 + 10} = \frac{10 \text{ mV}}{100 \text{ mV}} = 0.1 \qquad (3\text{-}2)$$

Hence we can select a value for either resistance arbitrarily and calculate the other. A few possible combinations are as follows:

R_1 (Ω)	R_2 (Ω)	I (A)
10	80	10^{-3}
100	890	10^{-4}
1000	9000	10^{-5}
10,000	90,000	10^{-6}

Note that for $R_1 + R_2 > 10^4$ Ω, the internal resistance of the GC can be overlooked, simplifying the calculation. If R_1 is greater than about 10^5 Ω, loading effects become noticeable. Hence $R_1 + R_2 = 10^4$ Ω would probably be the best choice in this example.

   An important use of voltage division occurs when an indicating meter is to be used for a number of ranges. Suppose a moving-coil meter is available with an inherent sensitivity of 1 V at 100 μA (often expressed as 10,000 Ω/V), and it is desired to provide ranges of 1, 5, 10, 50, and 100 volts, full-scale. This can be accomplished by means of a five-position range switch with corresponding resistors, as in Figure 3-8. For the 1-V range, no resistor is needed. For the 5-V range, the relation of Eq. (3-1) gives

$$\frac{E_{out}}{E_{in}} = \frac{1}{5} = \frac{10^4}{R_S + 10^4} \; ; \qquad R_S = 4 \times 10^4 \; \Omega \qquad (3\text{-}3)$$

The reader can verify the remaining values of R_S in the figure.

   If the same meter movement is to be used for current measurements, the range resistors must be connected in parallel (shunted) across the meter, as shown

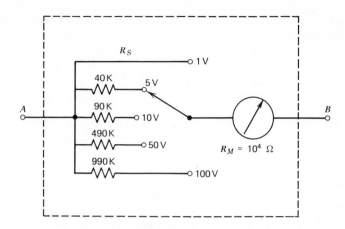

Figure 3-8.   A voltmeter with selectable ranges.

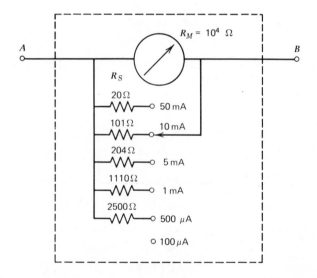

Figure 3-9.   An ammeter with selectable range resistors.   The type
of switch known as "make-before-break" must be used to protect
the meter from overload.

in Figure 3-9.   The parallel resistance of the meter
and shunt must always be such that the desired current
will produce a drop of 1 V (for full-scale deflection).

For example, the shunt for the 50 mA range would be calculated as

$$R_{total} = \frac{R_M R_S}{R_M + R_S} = \frac{10^4 R_S}{10^4 + R_S} \qquad (3\text{-}4)$$

But we also know that R_{total} = 1 V/0.05 A = 20 Ω, so that

$$R_S = \frac{20 \times 10^4}{10^4 - 20} \cong 20 \ \Omega \qquad (3\text{-}5)$$

The meter shunt is an excellent example of a *current divider*, as shown in Figure 3-10. One can readily prove that the following relation holds:

$$\frac{I_1}{I_2} = \frac{R_2}{R_1} \qquad (3\text{-}6)$$

The shunt values in Figure 3-9 can be calculated either from Eq. (3-5) or (3-6).

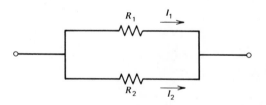

Figure 3-10.   A current-dividing circuit.

CAPACITORS

A conventional capacitor consists of two conductors made of metal foil separated by an insulating layer. There are three principal classes of capacitors: (1) *air-dielectric:* these are the variable capacitors widely used in tuning radiofrequency circuits; (2) *electrolytic:* aluminum or tantalum foil separated by a very thin layer of oxide produced electrolytically; and (3) *nonelectrolytic:* the metal foils are separated by layers of mica, paper, or various plastic films.

Capacitors are rated by their capacitance in farads (or more commonly microfarads, μF, or picofarads, pF), and by the maximum permissible voltage at which they can safely be used. The several types vary also in secondary characteristics, which may be of great practical importance.

Electrolytic units (unless marked "nonpolarized") must be connected in such a way that a DC voltage is maintained across them with the polarity marked on the case; wrong polarity may well cause explosive failure. These are especially useful for shunting out unwanted AC from a DC circuit, since they can be obtained with very large capacitance at low cost. Tantalum electrolytics provide the greatest capacitance per unit volume, but are generally more expensive than their aluminum counterparts. Electrolytic capacitors behave as if they had a resistor of a few megohms in parallel (leakage resistance), and this limits their field of application.

Of the nonelectrolytic types, those with dielectric of polystyrene or polycarbonate are to be preferred over paper in many applications, since their capacitance remains constant with time, and their leakage resistance is very high. Ceramic disc capacitors are often more convenient than the paper types, and of similar cost. Many types of capacitors, especially those made by winding long strips of foil into a cylindrical form, have considerable inductance. When it is important to pass a wide band of frequencies, it is good practice to wire a small mica or ceramic capacitor in parallel with any large paper or plastic film unit.

INDUCTORS

A simple inductor consists of a coil of wire wound around a core, which may be iron or ferrite,* and is characterized by inductance expressed in henries (H). The iron-core types have high inductance and are used principally in high-current power supplies and in re-

---

*Ferrite is a ceramic-like material made by sintering together various heavy-metal oxides. It has particularly favorable magnetic properties for the applications described.

lays. The ferrite inductors are important in radiofrequency equipment, for example, as antenna coils in
radio receivers. An inductor with two or more windings
is a *transformer*, useful for coupling information from
one circuit into another. Iron-core transformers are
used to change the voltage level of AC power lines.

Just as capacitors may have finite amounts of inductance, so inductors have some inherent capacitance,
which must be taken into account when high frequencies
are involved. Since inductors are made of wire, they
also possess DC resistance, which may be considerable.

IMPEDANCE

A network of (ideal) resistors shows the same
current-voltage relations when measured with AC as with
DC, but that is not true for circuits containing capacitors or inductors (*reactive components*). Therefore
a more general concept is needed to denote the ratio
of voltage across a component to the current flowing
through it, applicable to all types of components.
This concept is *impedance*, denoted by the letter Z,
and measured in ohms.

For a resistor there is no distinction between resistance and impedance (i.e., $Z_R = R$). For a capacitor, the impedance Z_C is given by

$$Z_C = \frac{1}{2\pi f C} = \frac{1}{\omega C} \tag{3-7}$$

where C is the capacitance in farads, f is the frequency in hertz, and $\omega = 2\pi f$ is the angular frequency
in radians per second. Thus a 1 μF capacitor measured
at 1 kHz will show an impedance of

$$Z_C = \frac{1}{(2\pi)\ (10^3)\ (10^{-6})} = 159\ \Omega \tag{3-8}$$

If the 1 kHz AC applied to this component is at 10 V,
the resulting current will be

$$I_C = \frac{E_C}{Z_C} = \frac{10}{159} = 62.9\ \text{mA} \tag{3-9}$$

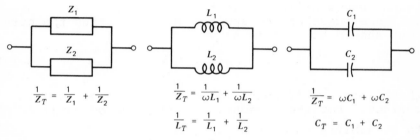

Figure 3-11. Series combinations of impedances; Z_T, L_T, and C_T are the respective total values.

$$\frac{1}{Z_T} = \frac{1}{Z_1} + \frac{1}{Z_2}$$

$$\frac{1}{Z_T} = \frac{1}{\omega L_1} + \frac{1}{\omega L_2}$$

$$\frac{1}{L_T} = \frac{1}{L_1} + \frac{1}{L_2}$$

$$\frac{1}{Z_T} = \omega C_1 + \omega C_2$$

$$C_T = C_1 + C_2$$

Figure 3-12. Parallel combinations of impedances.

The impedance of an inductor, Z_L, is given by

$$Z_L = 2\pi f L = \omega L \qquad (3\text{-}10)$$

where L is the inductance in henries.

Impedances of combinations of components follow the same rules for series and parallel connections given previously for resistors, as shown in Figures 3-11 and 3-12. Because of the reciprocal relation of Eq. (3-7), it can readily be seen that capacitances in parallel add, while in series they combine reciprocally. In contrast, inductances follow the same combination rules as resistances. This would be expected from the direct relationship of Eq. (3-10).

A different situation arises when various kinds of components are combined into a single circuit. Consider the series circuit of Figure 3-13. The applied AC excitation can be described as

$$E = E_0 \sin \omega t \qquad (3\text{-}11)$$

(a sine wave with amplitude E_0). It is reasonable to

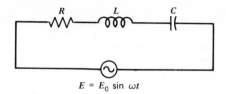

$$E = E_0 \sin \omega t$$

Figure 3-13. A series circuit containing resistance, inductance and capacitance, energized by AC at a frequency $f = \omega/2\pi$ Hz.

assume that the current that will flow in the circuit will also be AC and of the same frequency ω, but one cannot necessarily assume that the current will have the same phase as the applied voltage. Thus the current should be written as

$$I = I_0 \sin (\omega t + \phi) \qquad (3-12)$$

where ϕ is a difference in phase angle expressed in radians. For the present, I_0 and ϕ can be considered to be arbitrary constants.

The current-voltage relations for resistors, inductors, and capacitors, respectively, are known to be:

$$\text{Resistors} \quad E = RI \quad (\text{Ohm's law}) \qquad (3-13)$$

$$\text{Inductors} \quad E = L\frac{dI}{dt} \qquad (3-14)$$

$$\text{Capacitors} \quad E = \frac{1}{C} \int I \, dt \qquad (3-15)$$

Note that the equations for inductors and capacitors are inverse in form. Since the voltage supplied by the source in Figure 3-13 must be equal to the sum of the voltages over the components, we can write for the circuit

$$RI + L\frac{dI}{dt} + \frac{1}{C} \int I \, dt = E_0 \sin \omega t \qquad (3-16)$$

Since the current I in this equation is that of Eq. (3-12), we can substitute it into the above to give

$$RI_0 \sin (\omega t + \phi) + LI_0 \, \omega \cos (\omega t + \phi) - \frac{I_0}{C\omega} \cos (\omega t + \phi) = E_0 \sin \omega t \qquad (3-17)$$

This relation holds for all values of t. Hence it holds for the special case where $t = -\phi/\omega$, and hence $\omega t + \phi = 0$, and we can write

$$LI_0 \omega - \frac{I_0}{C\omega} = E_0 \sin (-\phi) = -E_0 \sin \phi \qquad (3\text{-}17)$$

(Remember that $\sin 0 = 0$ and $\cos 0 = 1$; see Appendix IV.)

For a second special case, let $\omega t + \phi = \pi/2$, from which $\sin (\omega t + \phi) = 1$ and $\cos (\omega t + \phi) = 0$, so that

$$RI_0 = E_0 \sin (\frac{\pi}{2} - \phi) = E_0 \cos \phi \qquad (3\text{-}18)$$

It is possible to eliminate both sine and cosine by squaring Eqs. (3-17) and (3-18) and adding them, since $\sin^2 \theta + \cos^2 \theta = 1$ by trigonometric identity. This gives us

$$E_0^2 = [R^2 + (L\omega - \frac{1}{C\omega})^2]I_0^2 \qquad (3\text{-}19)$$

We can determine the impedance of the complete RLC circuit, since the impedance Z is defined as the ratio E/I, by rewriting Eq. (3-19) in the form

$$Z = \frac{E_0}{I_0} = [R^2 + (L\omega - \frac{1}{C\omega})^2]^{1/2} \qquad (3\text{-}20)$$

It will be noticed that the right-hand side of this equation, the square-root of the sum of two squares, is identical in form to the well-known Pythagorean relation concerning the sides of a right triangle. Hence a geometrical diagram (Figure 3-14) can be used to determine the impedance of a combined RLC circuit. In this diagram, the resistance is represented by a horizontal vector pointing to the right, while the impedance of the inductor (called *inductive reactance*) is a vector pointing downward, and the *capacitive reactance* is a vector pointing upward. By the conventional rules for combining vectors, one can thus obtain the resultant Z. This procedure gives not only the magnitude of Z but also indicates a phase angle, defined trigono-

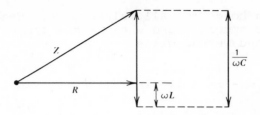

Figure 3-14. Vector diagram showing impedance relations for the circuit for Figure 3-13. By convention, $1/\omega C$ (capacitive reactance) is plotted upward, and ωL (inductive reactance) downward.

metrically. It might come as a surprise that an impedance in ohms has a phase angle associated with it, but this simply expresses the fact that the voltage and current are not in phase. For pure resistance, the phase angle is zero.

It should be noted that diagrams such as that of Figure 3-14 are commonly treated in terms of a complex plane. Reactances are taken as imaginary, i.e., ωC or ωL is multiplied by j (= $\sqrt{-1}$):

$$X_C = \frac{1}{j\omega C} = -j\frac{1}{\omega C} \quad \text{and} \quad X_L = j\omega L \qquad (3\text{-}21)$$

while resistance is taken as real. The net impedance, then, is in general a complex number, the exception occuring when $X_C = X_L$. A more thorough discussion appears in Chapter XII.

PHASE RELATIONS

Ohm's law applied to a pure resistance (no reactance) is independent of time. If the voltage increases, the current increases simultaneously. If a sine wave voltage is applied, the current will follow exactly. That is to say, the current is *in phase with* the potential.

In the case of a capacitance this is not so; and there is in fact a time difference between voltage and current. This can be made clear by considering what happens when the switch in the DC circuit of Figure 3-15 is closed. At this moment current starts to flow, charging the capacitor. The voltage across the capa-

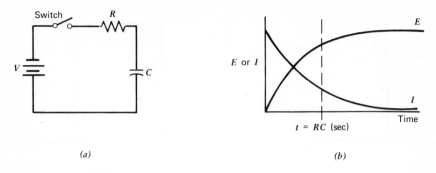

Figure 3-15. (a) DC charging circuit for a capacitor; (b) voltage
and current at the capacitor as functions of time.

citor, however, starts at zero and builds up slowly,
approaching asymptotically the battery voltage V, and
at the same time the current falls off toward zero.
Both reach 63% of their change of value in a time period equal to the product of R and C, in seconds. This
(RC) is called the *time constant* of the circuit. Since
the current flowing into the capacitor starts at a
large value and drops off, while the voltage across it
starts at zero and builds up, the current is said to
lead the voltage. If the battery is replaced by a
source of AC (Figure 3-16), then both current and voltage follow sine waves, but the phase of the current
will be found to *lead* the voltage by 1/4 cycle (i.e.,
90° or $\pi/2$ radians).

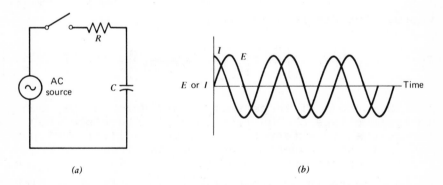

Figure 3-16. (a) An AC-driven *RC* circuit; (b) a plot of current
and voltage at the capacitor. The current and voltage scales
have been chosen to make the amplitudes appear equal.

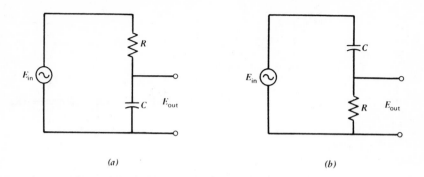

(a)   *(b)*

Figure 3-17. Two *RC* voltage dividers.

In an inductor, the effect is exactly reversed; the current in this case *lags* the voltage by 90°.

REACTIVE VOLTAGE DIVIDERS

It is instructive to investigate the properties of voltage dividers composed of resistors and capacitors (Figure 3-17). In Figure 3-17*a*, the output voltage is

$$E_{out} = E_{in}\left(\frac{Z_C}{R + Z_C}\right) = E_{in}\left[\frac{1/\omega C}{R + (1/\omega C)}\right] = E_{in}\left(\frac{1}{\omega RC + 1}\right)$$
$$(3-22)$$

whereas for Figure 3-17*b* it is

$$E_{out} = E_{in}\left(\frac{R}{R + Z_C}\right) = E_{in}\left[\frac{R}{R + (1/\omega C)}\right] = E_{in}\left(\frac{\omega RC}{\omega RC + 1}\right)$$
$$(3-23)$$

The output of these voltage dividers can be seen to be functions of the frequency ω. Therefore they can be used to attenuate selectively (filter out) portions of the frequency spectrum. In circuit *a*, if frequencies are low enough that $\omega RC \ll 1$, then $E_{out} = E_{in}$ and there is no attenuation. In contrast at high frequencies, $E_{out} \ll E_{in}$, and signals are filtered out. This is called a *low-pass filter*. In the second circuit, the effect is just the opposite: if the frequency

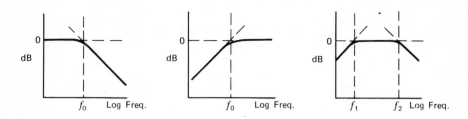

Figure 3-18. Response plots of passive filters: (a) low-pass, (b) high-pass, (c) band-pass.

is low, so that $\omega RC \ll 1$, then $E_{out} = E_{in}(\omega RC)$, which means that $E_{out} \ll E_{in}$, thus filtering out the low frequencies. On the other hand, if $\omega RC \gg 1$, then by Eq. (3-23), $E_{out} = E_{in}$. This constitutes a *high-pass filter*. The input-output relations as functions of frequency are shown in Figure 3-18. Note that the plots consist of **two** intersecting straight lines, somewhat rounded off. The intersection point is at the frequency $f_0 = 1/(2\pi RC)$, where the impedances of R and C are equal. Theory tells us that the rounded curve should lie below the intersection point by $20 \log \sqrt{2} \cong 3$ dB, so this is often called the "3-dB point."

A *band-pass filter*, in which both low and high frequencies are attenuated (Figure 3-18c) is also useful, but is not conveniently constructed with passive components alone.

Let us look now at some applications of these several types of filters. A high-pass filter is useful in improving the S/N ratio when information is modulated on an AC carrier. In this case, the $1/f$ noise and the 60-Hz interference can be filtered out without appreciably attenuating the information, provided that f_0 is properly chosen. An example is shown in Figure 3-19. Figure 3-19b shows the rationale for the choice of f_0. The value of 600 Hz was selected because it is 10 times the line frequency. Since the slope of the curve is 20 dB/decade, the attenuation at 60 Hz is 20 dB, a factor of 10. At the same time, 1000 Hz is sufficiently far removed from f_0 to be left unchanged by the filter. The upward slope of the $1/f$ noise is compensated by the opposite slope of the filter, and hence the noise is effectively removed.

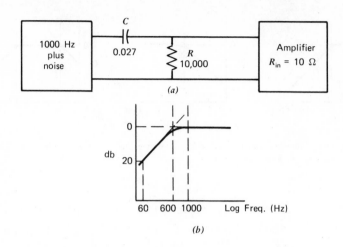

Figure 3-19.   (a) A high-pass filter coupling a 1000-Hz noisy
source to an amplifier.   (b) Corresponding Bode diagram;   at f_0
(600 Hz) the signal is attenuated to 71% (3 dB down).

The selection of a value for R (10,000 Ω) was made
on the same basis as for the example of Figure 3-7.
The value of C is calculated by the formula $C = 1/(2\pi$
$Rf_0)$, and rounded off to the nearest standard value
for capacitors.*   Since the frequency response curve
is smooth, a ±10% or even greater tolerance is acceptable.

In contrast, DC measurements can benefit from low-
pass filtration to attenuate noise, including AC interference.   The calculations are analogous to the
above, with interchange of R and C.   An example is
shown in Figure 3-20.

Another, more complex, type of filter is the *notch
filter*, shown in Figure 3-21.   This can be considered
to be a parallel combination of a modified low-pass
filter (R, R, $2C$) with a matching high-pass filter (C,
C, $R/2$).   The impedance of the network is very large
(in the megohm range) at the notch frequency, $f_0 = 1/$
$(2\pi RC)$, and low at other frequencies.   The sharpness of

*Only a selected sequence of values, the *standard values*, are
available commercially for both capacitors and resistors.

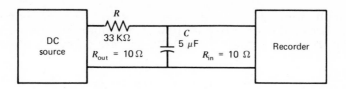

Figure 3-20. A low-pass filter with a time constant RC = 0.16 sec., meaning that of a sudden change (step) of input voltage, 63% will be recorded within 0.16 sec. This may slightly round off sharp peaks in the recorded signal. The value of f_0 is about 1 Hz.

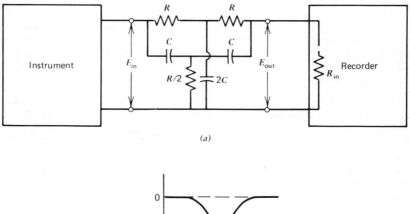

(a)

(b)

Figure 3-21. The Twin-Tee or notch filter. (a) An example of its use to couple the output of an instrument to a recorder with the elimination of interference at frequency f_0 = 1/($2\pi RC$). The value of R_{in} must be larger than R but smaller than the impedance of the filter at f_0. (b) Its frequency response.

the notch depends on the precision of the components. This filter, tuned to the power frequency, is widely used to remove traces of 60-Hz noise from DC circuits.

EQUIVALENT CIRCUITS

The circuits we have discussed thus far have been treated as black boxes, largely in terms of their transfer coefficients, E_{out}/E_{in}. This black-box designation is appropriate because it is possible to enclose a network within a box provided only with input and output binding posts, and to determine by external measurements alone all one needs to know about its electronics.

Thus, all one needs to know about an amplifier is indicated in Figure 3-22. The input and output impedances simply reflect the respective voltage-current relationships. The functioning of the amplifier is such that the input voltage E_{in} is somehow sensed, multiplied by the gain G, and fed to the output resistor. This representation gives no details about the actual internal circuitry of the amplifier. This is an important point because it facilitates interchangeability of a variety of modular amplifiers.

We shall make a more complete treatment of the simpler case of a device that has only two terminals rather than four as in the amplifier. Let us consider a cir-

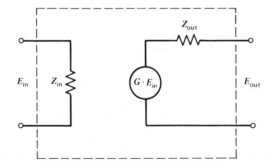

Figure 3-22. Representation of an amplifier. Not only Z_{in} and Z_{out}, but also the gain G, are functions of frequency.

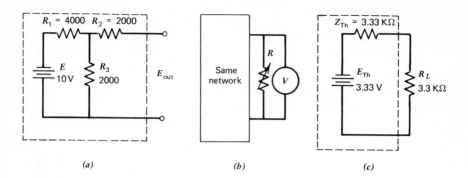

Figure 3-23. A black-box containing a battery and three resistors.
(a) Actual circuit. (b) Same, arranged for step-2 measurements.
(c) Thevenin equivalent circuit.

cuit box containing a 10-V battery and several resistors (Figure 3-23). If we measure the output voltage
with a high impedance voltmeter, we will obtain a reading of 3.33 V. (This is derived from the voltage-
divider equation; R_2 is ignored since only negligible
current passes through it.) Let us now connect a variable resistor R_L across the output while still measuring the voltage (Figure 3-23b). If R_L is large compared to the resistors inside the box (no loading) the
voltmeter will still read 3.33 V; at the other extreme,
if R_L is reduced to zero, the voltmeter will read zero.
So somewhere in between, we can locate a value of R_L
such that the voltage is reduced to exactly half of its
original value, namely 1.67 V. This will be found to
be 3.33 kΩ. We can conclude from this that the effective resistance inside the box is also 3.33 kΩ, because
(Figure 3-23c) only this gives a voltage divider with
two equal segments, cutting the voltage in half. Thus
the original network of three resistors and a 10-V battery can be replaced by a single resistor of 3.33 kΩ
and a battery of 3.33 V, and by external measurements
we could never detect the difference!
    This "gedanken experiment" can be generalized as
follows: *Any network made up solely of impedances and
voltage sources at a single frequency can be replaced
(as far as its behavior is concerned) by an equivalent
circuit consisting of one voltage source at the same*

frequency and one series impedance. This is the *The-venin theorem*, and the equivalent components are called the Thevenin voltage, E_{Th}, and the Thevenin impedance, Z_{Th}. This is a very useful concept in simplifying our thinking about many laboratory instruments that can be represented as a signal source acting through an output impedance.

The Thevenin quantities for known circuits can be calculated as follows: (1) By suitable application of Ohm's law, ascertain what the output voltage would be with no current flowing; this is E_{Th}. (2) Calculate what *current*, I_{sc}, would flow through the output ter-minals if they were tied together (short-circuited). The ratio E_{Th}/I_{sc} gives the Thevenin impedance. As an example, consider again the circuit of Figure 23*a*. For step (1), R_2 can be ignored, and E_{out} is given by the voltage-division equation

$$E_{out} = \frac{R_3}{R_1 + R_3} \cdot E_{in} = 3.33 \text{ V.} \qquad (3\text{-}24)$$

For step (2), the output is short-circuited, and R_2 combined in parallel with R_3 to give 1000 Ω. In turn, R_1 is added, to give 5000 Ω. The total current is therefore 2 mA, and the current through the short cir-cuiting wire, I_{sc}, is half of it, or 1 mA. Consequent-ly the Thevenin quantities are

$$E_{Th} = 3.33 \text{ V} \quad \text{and} \quad Z_{Th} = \frac{3.33 \text{ V}}{0.001 \text{ A}} = 3.33 \text{ k}\Omega$$

IMPEDANCE MATCHING

When a signal is transmitted from a network to its load, inevitably some of the power is dissipated inter-nally and some in the load. In some cases it may be useful to optimize the system in such a way that max-imum power is transmitted to the load. Such is the case in power systems like radio transmitters, but it is also important at very low power levels, in order to optimize the *S/N*. In Figure 3-24 is shown an example.

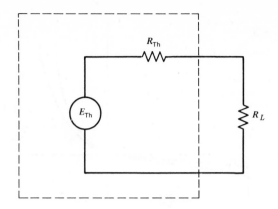

Figure 3-24. Circuit illustrating the transfer of power from source to load.

The power in the load is calculated by

$$P = I^2 R_L \tag{3-25}$$

but

$$I = \frac{E_{Th}}{R_{Th} + R_L} \tag{3-26}$$

Therefore

$$P = E_{Th}^2 \cdot \frac{R_L}{(R_{Th} + R_L)^2} \tag{3-27}$$

It is left to the reader to show that when R_L is very large, the power transfer reaches a maximum for $R_L = R_{Th}$.

For most applications, however, it is advantageous instead of maximizing the transfer of power, to minimize the effect of impedance *changes*. Variations in either R_L or R_{Th} will have small effects on E_{out} when $R_L \gg R_{Th}$. This is the situation normally encountered in instrumentation systems. (See, for example, Figure 3-19.)

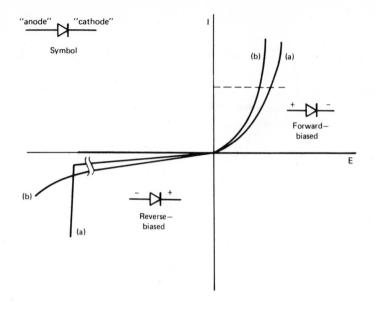

Figure 3-25.   Characteristic curves for semiconductor diodes.   (*a*)
Silicon.   (*b*) Germanium.

DIODES

A diode is a two-terminal device that passes current easily in one direction (forward) and not in the
reverse direction. Its resistance therefore is widely
different when measured with one polarity or the other.
A typical unit may have a resistance of 10 Ω in the
forward direction and 100 MΩ in reverse. The conventional symbol and representative current-voltage curves
are shown in Figure 3-25. The direction of the arrow-
head indicates the direction of easy flow of conventional positive charges.

The residual current that flows when a diode is
reverse-biased is very small, measured in microamperes
or nanoamperes. The potential drop across the diode
during forward conduction (*forward voltage drop*; note
the horizontal dashed line) is, at a given current,

approximately twice as large for a silicon as for a
germanium diode. At higher current values, the forward
drop becomes fairly constant, about 0.3 to 0.4 V for
germanium and 0.6 to 0.8 V for silicon.

The primary function of a diode is to allow current
to pass in one direction only. Accordingly one of its
major uses is in rectification: the conversion of alternating to direct current. Many applications of
diodes are described in later chapters.

One common application of a diode is as a *clamp* or
"DC-restorer." Suppose that a square wave with 10-V
peak-to-peak amplitude is passed through a capacitor,
as in a high-pass filter (Figure 3-26). If the *RC*-
time constant is long compared to the period of the
wave, the shape of the wave will not be impaired by the
capacitor, but the wave will no longer have a predict-
able relation to ground. This relation can be reestab-
lished by the diode clamp which prevents the signal
from becoming negative. The entire 10-V swing is then
forced to be positive with respect to ground. Allow-
ance must be made for the forward voltage drop in the
diode, so that (for germanium) the square wave shown
would actually swing from -0.3 to +9.7 V. Hence this
kind of clamp is not applicable to very small signals.

Another class of diode application depends on the
very nearly logarithmic shape of the current-voltage
curve of a silicon diode in the first quadrant (curve
a of Figure 3-25). This permits multiplication and division of electrical quantities by logarithmic transformation, considered in a later chapter.

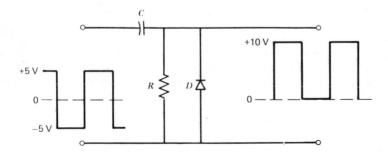

Figure 3-26. A diode clamp.

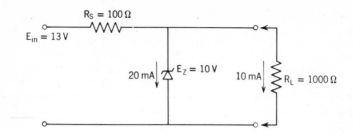

Figure 3-27.   Circuit illustrating the use of a zener diode.

ZENER DIODES

Returning to Figure 3-25, observe that both curves drop off more or less sharply when the reverse voltage is increased sufficiently. This limits the amplitude of the AC voltages that can be rectified. It is possible, however, to make good use of this reverse voltage limitation in silicon, where the drop-off is very sharp. A diode designed for reverse applications is called a *zener diode* or *breakdown diode,* and the reverse voltage that will cause significant current to flow is the zener or breakdown voltage.

Zener diodes are widely applied to voltage regulation. An elementary circuit is shown in Figure 3-27. (Note that the symbol for a zener diode is that of a normal diode with wings added, reminiscent of a Z.) The objective of the circuit is to maintain a constant voltage across the load resistor, R_L, in this case 10 V, regardless of variations in both E_{in} and R_L. In the absence of the zener, the circuit will act as a voltage divider and the objective will not in general be attained. With the zener in place, the system still acts as a voltage divider as long as the value at A is too small, but when the voltage across the zener diode reaches 10 V, it triggers the zener process and clamps at exactly 10 V for larger values of E_{in}. Variations of R_L will likewise have no effect provided that the voltage at A calculated by the voltage-divider equation is larger than 10 V. For the values indicated in the figure, the difference between E_{in} and 10 V appears across R_s, and the current is given by $(E_{in} - 10)/R_s = 3/100 = 30$ mA, of which 10 mA will be used by the load and 20 mA by the zener.

As is always the case, the power ratings of the various components must not be exceeded. In this example, R_L (1 kΩ) need only be capable of dissipating $P = (0.01)^2 \times 1000 = 0.1$ W. If E_{in} might go as high as 15 V, R_s should be rated at 0.25 W, and the zener at a minimum of 0.5 W. In laboratory-made instruments, it is highly advisable that one select components of at least twice the calculated power rating to allow for poor heat dissipation and to make the circuit more robust.

The temperature coefficient of the zener voltage, although small, can sometimes be significant. For E_z values of approximately 5 to 7 V the coefficient is nearly zero; for higher values it is positive, for lower, negative. The *forward* voltage drop of a silicon diode has a negative temperature coefficient, so that it is often possible to combine one or several forward diodes with a reversed zener (all at the same temperature) to produce a temperature-compensated voltage reference. For example, three 1N536 diodes in series with a 1N1604 zener gives a combined voltage drop of approximately 12.2 V with a temperature coefficient of only 0.004%/°C. The temperature-compensated zeners available commercially consist of this kind of combination; they often have less sharp zener limitation than usual.

Zener diodes are useful in protecting various electronic components against excessive voltage. For example, a 10-V zener in series with a small resistor, mounted across the terminals of a 10-V DC meter will prevent damage from higher voltage of the correct polarity or from any reverse voltage that might be applied. Zeners can similarly protect switch contacts from arcing, as when an inductive load is suddenly disconnected (Figure 3-28). A pair of matched zeners back-to-back form a symmetrical clipping element (Figure 3-29); if the amplitude of the AC is large compared to the zener voltage, the resulting truncated sine wave will be a close approximation to a square wave.

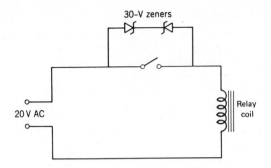

Figure 3-28. An AC relay circuit, with the switch protected by a pair of zeners. In such a pair of back-to-back zeners, for either polarity, one diode operates in the zener region, the other being forward biased.

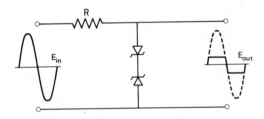

Figure 3-29. Clipping circuit using zener diodes.

PROBLEMS

3-A.   Determine the total impedance of a parallel combination of ten 1-MΩ resistors.

3-B.   Consider the circuit of Figure 3-6b.   The input is maintained at a fixed voltage, E_{in}.

(a)   Compute E_{out}, I_{load}, and I_{in}.

(b)   From the quantities in (a), compute the following:

Input impedance $Z_{in} = \dfrac{E_{in}}{I_{in}}$

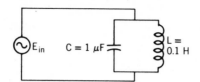

Figure 3-30. See Problem 3-D.

$$\text{Voltage gain} \quad A_v = \frac{E_{out}}{E_{in}}$$

$$\text{Current gain} \quad A_I = \frac{I_{load}}{I_{in}}$$

3-C.   Consider again the network of Figure 6-3b, with a 10 V source at the input.

(a)   Compute the voltage gain in decibels.
(b)   Compute the current gain in decibels.

3-D.   For the circuit of Figure 3-30, there is a frequency f_R at which $\omega L = 1/\omega C$. Compute the impedance at that frequency. This is known as the *resonant frequency* of the LC circuit and is given by $f_R = 1/(2\pi\sqrt{LC}\,)$.

3-E.   An AC signal of 150 V RMS at 10,000 rad/sec is applied to a capacitor of 100 pF.

(a)   Compute the current and the power dissipation.
(b)   What is the frequency of the signal in kilohertz?

3-F.   For the circuits of Figure 3-31, calculate the voltages across and the current through the load, for R_L = 100, 1000, and 10,000 Ω. Show the difference between current and voltage sources to be simply a matter of impedance ratios.

3-G.   Repeat the procedure of Problem 3-F with the circuits of Figure 3-32.

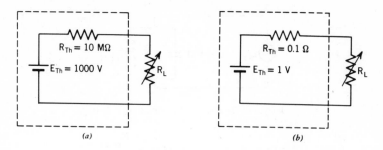

Figure 3-31.   See Problem 3-F

3-H.   Calculate a low-pass filter of 1 kΩ input inped-
       ance and f_0 = 500 Hz.

3-I.   A ±1 V zero-center meter with internal resist-
       ance of 10 kΩ, for use as a null indicator, can
       be provided with a pair of shunt diodes. This
       not only protects the meter from overloads, but
       also considerably extends the useful range while
       retaining its sensitivity near the null point.
       Explain.

                    *   *   *

3-1.   What is the total impedance, Z_t, of a parallel
       combination of Z_1 = 100 Ω and Z_2 = 151 Ω?
       Explain.

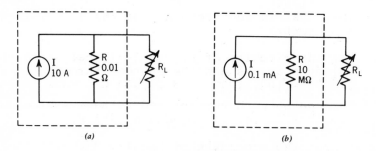

Figure 3-32.   See Problem 3-G.

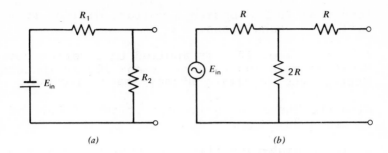

Figure 3-33.   See Problems 3-4 and 3-6.

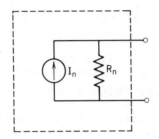

Figure 3-34.   See Problem 3-7.

3-2.   A voltage divider is desired to attenuate a 10-V
       signal   to 10 mV.   In addition, the current from
       the source must not exceed 1 mA, and R_{out} must be

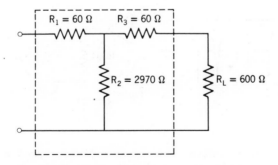

Figure 3-35.   See Problems 3-8 and 3-9.

less than 20 Ω. Design a voltage divider to fulfill these conditions.

3-3.   A DC signal of 10 V is applied to a series combination of a 100 Ω resistor and 100 mH inductor. What is the resulting steady-state current?

3-4.   Find the Thevenin equivalent of the circuit of Figure 3-33a.

3-5.   Design a high-pass filter to operate from an AC source. The frequency f_0 is to be such that $\omega = 10^6$ rad/sec.

3-6.   Find the Thevenin equivalent of the circuit of Figure 3-33b.

3-7.   Consider the circuit consisting of an ideal current source and a parallel resistor, R_n, as shown in Figure 3-34. Find its Thevenin equivalent.

3-8.   (a) In the circuit of Figure 3-35, with the load connected, compute the input resistance and also the voltage, current, and power attenuations (expressed in dB). (b) What will happen to the input impedance if a number of similar "black boxes" are connected in series between source and load?

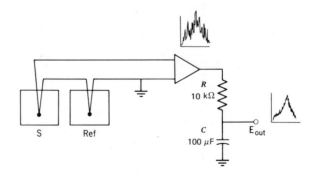

Figure 3-36.   An RC-filter for noise rejection used with a thermocouple.   The S represents the object under study, Ref is at a reference temperature (e.g., 0°C).

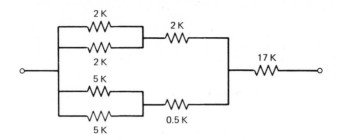

Figure 3-37.   See Problem 3-11.

3-9.   Consider a low-pass filter with attenuation slope
        of 40 dB/decade and f_o = 3000 Hz.   The input
        consists of a composite signal containing a fun-
        damental frequency of 2000 Hz (10 V), with 2 V
        of second harmonic and 1 V of third harmonic.
        Compute the harmonic content of the output.

3-10.   Consider the circuit of Figure 3-36.   The signal
        is 100 mV DC, and 10 mV of 60-Hz noise is pre-
        sent.   If R = 10 kΩ and C = 100 μF, what is the
        S/N ratio improvement?

3-11.   Calculate the equivalent resistance of the cir-
        cuit in Figure 3-37.

3-12.   Diodes and dry batteries connected as in Figure
        3-38 form a useful clipping circuit.   What would
        be the value of E_{out} to be expected from each of
        the following input functions: (a) E_{in} = +1.0 V;
        (b) E_{in} = -10.0 V; (c) E_{in} = 10 sin ωt, where ω
        = 100 rad/sec; (d) E_{in} = (10 - 2t) V.   Sketch
        the outputs for (c) and (d) as functions of time.
        Neglect the forward voltage drops of the diodes.

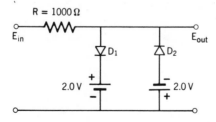

Figure 3-38.   See Problem 3-12.

IV

OPERATIONAL AMPLIFIERS

An operational amplifier (OA or op amp) is a modular unit, symbolized by a triangle (Figure 4-1), and characterized by high-voltage gain, high-input impedance and low-output impedance. A typical unit is limited to perhaps 10 mA current and ±10 V output.* In addition it is zero-crossing, meaning that its output can swing in both positive and negative senses, with zero volts input giving zero volts output. It is provided with two input connections (a and b in Figure 4-1), and a single output. The two input voltages subtract, so that the output is given by

$$E_{out} = A(E_b - E_a) \qquad (4\text{-}1)$$

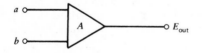

Figure 4-1. General symbol for an operational amplifier.

*Many commercial amplifiers are guaranteed to give at least a 10-V output, but actually will go to 11 or 12 V before saturating.

This expression leads to several significant con-
clusions. In the first place, since the gain is high,
the permissible difference between inputs in limited to
a very small value. Suppose, for example, that the
gain $A = 10^6$, then, since the output cannot exceed 10.
V, it follows that the differential input must not be
greater than 10 μV. A greater differential input
causes the amplifier to go into saturation at somewhat
above 10 V, and the equation is no longer obeyed.

Furthermore, if one input is grounded (i.e., at
zero volts), the output will follow the sign predicted
by Eq. (4-1). Thus if a is grounded, the output will
be of the same sign as E_b, whereas if b is at ground,
the output will take the sign opposite to E_a. Because
of this property, b is called the *non-inverting* input,
while a is the *inverting* input.

As a result, an amplifier can be used as a *comparator*, to give an indication of the relationship between
two voltages: If $E_a > E_b$, the amplifier will show -11
V; if $E_a < E_b$, the output will be +11 V, if this is
the saturation level. Note that this mode of operation
can be classified as digital, since it permits only
two states. A comparator finds many applications. For
example, it can energize a relay to terminate an operation when some variable exceeds a preset value. We
will return later to a more detailed discussion.

The comparator is a nonlinear device. Many more
applications are possible if the amplifier is made to
operate in a linear mode, with an analog rather than
digital output. This can be accomplished by incorpor-
ating *negative feedback*, a signal path between the output and the inverting input, which is best explained
through an example. Consider the amplifier in Figure
4-2, with the noninverting input at zero potential
(grounded), and resistors connected from the inverting
input both to the signal (E_{in}) and to the output. We
know from Ohm's law that the current flowing through a
resistor is given by the difference of potential between its terminals divided by its resistance. Therefore the input current $I_{in} = (E_{in} - E_{SJ})/R_{in}$ and the
current I_f, called the *feedback current*, is ($E_{SJ} -
E_{out})/R_f$. Now the input impedance of the amplifier is

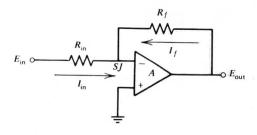

Figure 4-2. An amplifier with negative feedback. The inverting input is called the *summing junction, SJ,* as the two currents, I_{in} and I_f, are said to be "summed" at this point.

very high, so that essentially no current is lost into the amplifier, and therefore I_{in} and I_f must be equal, so we can write

$$\frac{E_{in} - E_{SJ}}{R_{in}} = \frac{E_{SJ} - E_{out}}{R_f} \qquad (4-2)$$

We also know that $E_{out} = A(E_b - E_a) = A(0 - E_{SJ})$, or $E_{SJ} = -E_{out}/A$. By substitution into both sides of Eq. (4-2) we obtain

$$\frac{E_{in} + \frac{1}{A}\cdot E_{out}}{R_{in}} = \frac{-\frac{1}{A}\cdot E_{out} - E_{out}}{R_f} \qquad (4-3)$$

which simplifies to

$$\frac{E_{out}}{E_{in}} = -\frac{AR_f}{(A + 1)R_{in} + R_f} \qquad (4-4)$$

This is the fundamental equation for this general circuit. It can be further simplified by using the relation that $(A + 1)R_{in} \gg R_f$, almost universally the case. Under these conditions,

$$\frac{E_{out}}{E_{in}} = -\frac{R_f}{R_{in}} \qquad (4-5)$$

This very important relation is valid only for very
large values of A and Z_{in}, both of which are properties
of the amplifier itself, and for $AR_{in} >> R_f$, which is
dependent on external circuitry. An amplifier that
meets these conditions is properly called an *operational amplifier*.

The gain that can be obtained from an op amp is
shown by Eq. (4-5) to be determined *only* by the external circuit. Thus if R_{in} = 10 kΩ and R_f = 100 kΩ,
the gain will be $-R_f/R_{in}$ = -10, regardless of the specific amplifier employed. It is of interest to compute the error that results from using Eq. (4-5) in
place of the exact equation (4-4), that includes the
gain A. It can readily be ascertained that for $A =$
10^6 and the resistance values given above, the devia-
tion is only 0.001%, or 1 part in 10^5. Hence the normal variation in A between different amplifiers will
have no significant effect on the circuit gain.

At this point a word is in order about the *ground*
concept. It is customary in electronic circuits to
show ground symbols at various points, as, for example,
at the noninverting input of the amplifier of Figure
4-2. This is intended to indicate the zero point   for
all voltages in the system. In practice all ground
points are connected together and to the ground terminal (common) of the power supply. Whenever voltages
are specified, it is to be understood that they are
taken relative to this ground.

For most zero-crossing amplifiers, a dual power-
supply is required, giving, typically +15 and -15 V,
with a common ground point. The connections between
power supply and amplifier are usually not shown explicitly, to avoid confusion.

The most useful feature of op amps, and the cause
of their prominent position in instrumentation, is
that they follow Eq. (4-5) and similar relations to a
remarkable degree of precision. The small residual
deviations are treated later in this chapter.

SINGLE-INPUT CIRCUITS

If the noninverting input is connected to ground,
the amplifier is said to be operating in the *grounded-
reference* or *single-ended* mode. In contrast, the

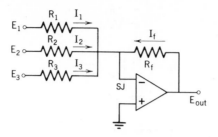

Figure 4-3. A voltage summer.

differential mode appears if both inputs carry signals.
(Grounding the inverting input is most unusual.)   In
single-ended operation, the summing junction is at
virtual ground, meaning extremely close to ground potential.

An example of a grounded-reference circuit is giv-
en in Figure 4-3. Assuming, as before, no loss of
current into the amplifier, one can write

$$I_1 + I_2 + I_3 = -I_f \qquad (4-6)$$

where the minus sign reflects the fact that the I_f arrow in the figure points in the opposite direction to
the others. The currents can then be expressed in
terms of the corresponding voltages, keeping in mind
that the summing junction is at virtual ground

$$\frac{E_1}{R_1} + \frac{E_2}{R_2} + \frac{E_3}{R_3} = \frac{E_{out}}{R_f} \qquad (4-7)$$

This can be rewritten in the form

$$E_{out} = -\left(E_1 \cdot \frac{R_f}{R_1} + E_2 \cdot \frac{R_f}{R_2} + E_3 \cdot \frac{R_f}{R_3} \right) \qquad (4-8)$$

Consequently this circuit sums all the input voltages,
multiplied by constant coefficients, with the usual
sign inversion. This circuit is called a *summer*. If
all the resistors are equal, the operation is the simple addition of voltages. If they are not equal, the

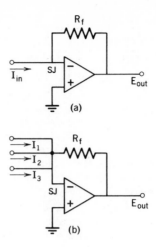

(a)

(b)

Figure 4-4. Current-to-voltage converters: (a) single-input, and (b) multiple-input, a current summer.

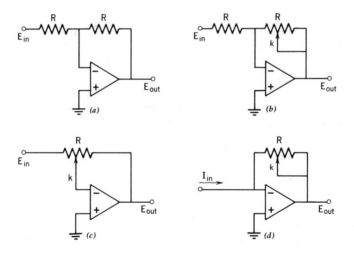

Figure 4-5. Examples of summer applications. The coefficient k is the fractional resistance exhibited by the section between the variable contact and the output. The circuit in (b) is linear with respect to k, and is usually preferred to (c).

circuit becomes a *weighted summer*. The voltages are, in general, time-dependent, whereas the weighting coefficients are constants. Multiplication by variable quantities requires considerably more complicated instrumentation.

As Eq. 4-6 indicates, the basic response of the amplifier and its network is to currents rather than voltages. The input resistors simply serve to establish the currents to be drawn from their respective signal voltages. If the desired information is already carried by a current, then an input resistor is unnecessary, and a direct connection to the summing junction can be made, as in Figure 4-4. This is a useful configuration in an instrument in which small currents must be measured.

Figure 4-4*b* shows a current summer, where the feedback current equals $I_1 + I_2 + I_3$. The output, by Ohm's law, is

$$E_{out} = -R_f \, (I_1 + I_2 + I_3) \qquad (4-9)$$

The negative sign indicates that if the flow of positive charges at the input follows the arrows, then the output is negative.

This circuit is called a *current-to-voltage converter*, in that it permits one to measure the signal as a voltage rather than a current. Thus, for $R_f = 10$ MΩ, if the input is 0.1 μA, the output will be 1 V, a quantity much easier to measure with an inexpensive meter than a very small current. An additional advantage is that the drop in voltage at the input as the current is fed in is essentially zero, since the summing junction is held at virtual ground. This is equivalent to zero input impedance,* an essential feature for exact current measurements. Also as a consequence of the virtual ground, several inputs will not interact with each other, all currents flowing to a common point of zero potential.

Voltage and current summers can be modified in a variety of ways, a few of which are shown in Figure 4-5.

---

*This might appear to be a contradiction, but remember that this is the impedance of the *circuit*, not of the amplifier per se.

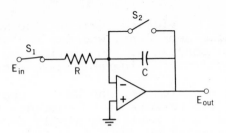

Figure 4-6. An analog integrator.

Most of these circuits are self-explanatory; the circuit in (c) permits a wide range of gains, essentially
from zero to infinity.
    Besides summing, there are two more operations
that can be carried out with almost equal simplicity:
integration and differentiation. These are implemented
by connecting capacitors in the feedback or input, respectively. (In principle, inductors could be used for
the same purposes, but in practice capacitors are always chosen.)
    The circuit of an *integrator* is shown in Figure
4-6. The current flowing in the feedback loop charges
the capacitor; its voltage, $E_{out} - E_{SJ}$, will therefore vary according to the equation

$$I_f \;=\; C \frac{dE_{out}}{dt} \qquad\qquad (4\text{-}10)$$

This current must equal the input current, so that

$$\frac{E_{in}}{R} \;=\; - \, C \frac{dE_{out}}{dt} \qquad\qquad (4\text{-}11)$$

This can be rearranged and integrated to give

$$E_{out} \;=\; - \frac{1}{RC} \int E_{in} \; dt + E_0 \qquad\qquad (4\text{-}12)$$

where E_0, the integration constant, represents the output voltage at time zero. Thus the output becomes the
time integral of the input. The product RC has the dimensions of time, and is called the *time constant*.
    The integrator can be used to generate a DC *ramp*

voltage, $E = kt$ (k being a constant), simply by holding E_{in} at some constant value. Then Eq. (4-12) becomes

$$E_{out} = - \frac{E_{in}}{RC} \int dt + E_0 = E_0 + kt \qquad (4-13)$$

The output of the amplifier starts at E_0 and increases at a constant rate until it reaches its maximum voltage (saturation). It is therefore necessary to provide some means of discharging the capacitor to reset the integrator to zero, ready for the next integration. The switch S_2 in Figure 4-6 serves this purpose. On the other hand, switch S_1, when opened, stops the integration, and the capacitor retains whatever charge it has at that moment, thus holding E_{out} at a constant voltage. The three modes of operation are called, respectively *integrate*, *reset*, and *hold*.

Also of interest is the integration of a sine wave to give a cosine wave

$$E_{out} = - \frac{1}{RC} \int E_0 \sin \omega t = + \frac{E_0}{\omega RC} \cos \omega t \qquad (4-14)$$

Note that the output amplitude is a decreasing function of frequency, and is equal to the input only for $\omega = 1/RC$.

An important application of integration is in *data processing*, meaning the modification of a signal to meet the requirements of the particular measurement. For instance, if the area under a recorded peak, rather than the peak height, represents the quantity of interest, then an integrator is always needed. This is shown in Figure 4-7. The height of the second wave in Figure 4-7*b* represents

$$H_2 = k \int_{t_1}^{t_2} E \, dt \qquad (41-15)$$

which represents the area under the second peak in Figure 4-7*a*, and where k is a negative constant.

Figure 4-8*a* shows a *differentiator*. The equation

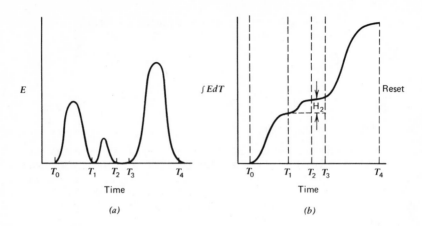

Figure 4-7. Example of peak integration, such as may be useful in chromatography or NMR spectroscopy.

describing its behavior is

$$C \frac{dE_{in}}{dt} = - \frac{E_{out}}{R} \qquad (4-16)$$

which can be rewritten as

$$E_{out} = - RC \frac{dE_{in}}{dt} \qquad (4-17)$$

In Figure 4-8a and b shows the results of differentiating a triangular wave and a sine wave, respectively. These can be rationalized if one remembers that the derivative is a measure of the slope of a function. Note that integration produces the opposite functional change, transforming the lower curves into the upper ones. In both cases the RC time constants effect the relative amplitudes of the curves. As a rule integration, rather than differentiation, is to be selected whenever possible because of its inherent enhancement of the S/N ratio.

Figure 4-8c shows that the operations of summing and differentiation can be combined in a single amplifier. The output is the sum of the derivatives of the two inputs.

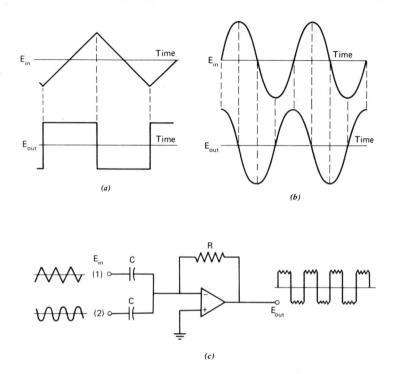

Figure 4-8. (a) The derivative of a triangular wave is a square wave. (b) The derivative of a sine is a cosine. (c) A summing differentiator.

The relations between various frequencies can be better understood through the following considerations. Any periodic signal, of fundamental repetition rate, f, can be expressed as a sum of sines and cosines, multiplied by appropriate constant coefficients, with the possible addition of a DC term. The arguments of the sines and cosines are $2\pi f$ and its multiples $2\pi n f$ (Fourier expansion). If the gain of a circuit is frequency dependent, the coefficients of the terms of the Fourier expansion are altered, and a *new* function is produced. Therefore, for such circuits, a gain as such cannot be defined for periodic functions, except for pure, single-frequency sine waves. As an example, the derivative and integral of a function are in general new functions, so that "gain" for a differentiator or inte-

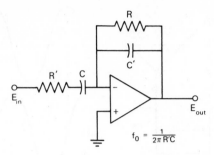

Figure 4-9. Practical differentiator. Differentiation occurs according to the formula $E_{out} = RC(dE_{in}/dt)$ between DC and f_0.

grator has no meaning. In the specific case of a pure sine or cosine wave, a gain can be defined. To deal with more complex cases a generalized quantity, the *transfer function*,* must be introduced, which fills a role similar to the gain. The transfer function is defined in terms of Laplace transforms, which are discussed in Chapter 12.

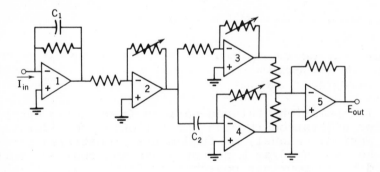

Figure 4-10. Circuitry for adding a derivative to the original signal. Amplifier 1 is a current-to-voltage converter (required for the particular application) in which capacitor C_1, a few picofarads, provides high-frequency attenuation. Amplifiers 2 and 3 are variable-gain amplifiers and amplifier 4 generates the derivative, which is added to the original signal at the summing junction of amplifier 5.

*The transfer function is not to be confused with the transfer coefficient, previously introduced.

For sine waves, the gain of the differentiator approaches zero at low frequencies and the inherent gain
A at high frequencies. This increase in gain is de-
trimental, in that noise at higher frequencies is
greatly amplified. For most purposes, therefore, the
differentiator must include a filter, which can be implemented as shown in Figure 4-9.
Combinations of summers, differentiators, and integrators can be used to generate sums and differences
of derivatives and integrals. An example of such a
combination, which produces the sum of an original
function and its derivative, is shown in Figure 4-10;
it has been used for improving the quality of pictures
obtained from an electron probe microanalyzer.*

DIFFERENTIAL MODE CIRCUITS

If both inputs of an op amp carry signals, the virtual ground condition is replaced by the virtual equality of the two inputs. Such differential operation is
more versatile than the single-ended mode. As an example, in Figure 4-11 is shown a subtractor. The output can be calculated in the following way. Using the
voltage divider equation, the voltage E_2', at the non-

inverting input, is given by

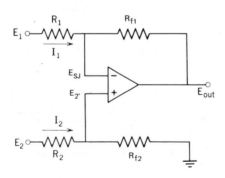

Figure 4-11. A circuit for subtraction of one voltage from another.

*K. F. Heinrich, C. Fiori, and H. Yakowitz, *Science,* <u>67</u>, 1129 (1970).

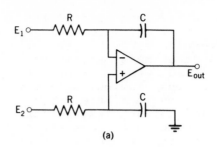

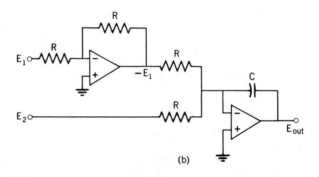

Figure 4-12. Difference integrators: (a) with a differential amplifier, and (b) a single-ended equivalent.

$$E_2' = E_2 \frac{R_{f2}}{R_2 + R_{f2}} \tag{4-18}$$

The value of E_{SJ}, as discussed above, must equal E_2', so that the current I_1 will be

$$I_1 = \frac{E_1 - E_{SJ}}{R_1} = \frac{E_1 - E_2'}{R_1} \tag{4-19}$$

Thus the equality between the input and feedback currents can be written as

$$\frac{E_1 - E_2'}{R_1} = \frac{E_{out} - E_2'}{R_{f1}} \tag{4-20}$$

By combining Eqs. (4-19) and (4-21) and taking $R_{f1} = R_{f2} = R_f$, and $R_1 = R_2 = R_{in}$, it follows that

$$E_{out} = \frac{R_f}{R_{in}} (E_2 - E_1)$$ (4-21)

Mathematically, the circuit takes the difference between two voltages and multiplies it by a constant; in electronic terms, it amplifies the difference between two voltages by the factor R_f/R_{in}. The same operation could be performed by single-ended amplifiers, but two would be required instead of one.

Along the same line, one can implement *difference integration*, as indicated in Figure 4-12*a*. In this case the output is given by

$$E_{out} = \frac{1}{RC} \int (E_2 - E_1) \, dt$$ (4-22)

Again it is possible to use single-ended amplifiers, as indicated in Figure 4-12*b*. The latter alternative is much to be preferred, since it requires only a single high-quality capacitor. The price of a second, matched, capacitor is likely to exceed the price of a second amplifier. As a matter of fact, most difference operations can be implemented equally well either with pairs of single-ended amplifiers or differentially.

A special case where a differential amplifier is restricted to single-ended operation is the *voltage follower*. The follower (Figure 4-13*a*) uses the simplest form of feedback, a direct connection between the output and the summing junction. The voltage gain of this circuit is unity and noninverting. In other words, it reproduces the input voltage. Its great merit is that it provides *impedance transformation*: the input impedance is very large (typically 100 MΩ), whereas the effective output impedance is a small fraction of an ohm. This corresponds to changing the Thevenin impedance from megohms to milliohms (Figure 4-13*b*). This impedance transformation is very useful in a variety of applications. One example is shown in Figure 4-14, where the follower is interposed between

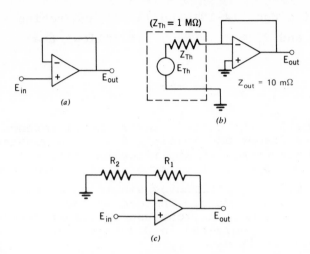

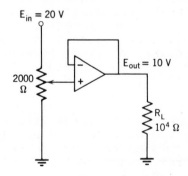

(c)

Figure 4-13. (a) A voltage follower. (b) The follower connected
to a source represented by its Thevenin Equivalent. (c) A follower with gain.

a potentiometer and its load to avoid the loading effects previously described (see Figure 3-6). This use
of a follower also reduces the noise often produced by
poor mechanical contact in a potentiometer;   any variation in the resistance of the sliding contact will be
totally negligible because no current will be passed
through it.

Figure 4-14. Unloading (buffering) of a potentiometer through
the use of a follower. Without the follower, the loading error
could amount to several percent.

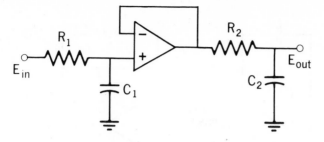

Figure 4-15. Buffering in a two-stage filter.

A voltage follower can be provided with gain by means of the feedback circuit of Figure 4-13c. The governing relation is

$$E_{out} = E_{in}\left(\frac{R_1}{R_2} + 1\right) \qquad (4\text{-}23)$$

Note that gains less than unity are not possible.

Another example of buffering is shown in Figure 4-15, where a two-stage RC filter is synthesized. In many such cases it is advantageous to separate the successive stages of filtering by followers.

Buffering can be accomplished with grounded-reference amplifiers if the elements to be isolated do not require that one terminal be connected to ground. As an example, one can use the circuit of Figure 4-16a to buffer a Weston standard cell (a widely-used voltage reference cell of approximately 1.0183 V). The out-

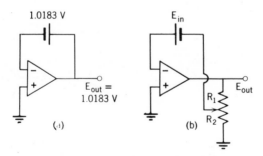

Figure 4-16. (a) Single-ended follower circuit used in conjunction with a Weston Standard Cell. (b) Single-ended follower with gain. The output is given by $E_{out} = [(R_1 + R_2)/R_2]E_{in}$.

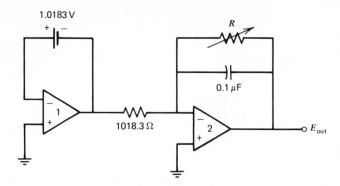

Figure 4-17. A high-quality voltage source. R is a 10-kΩ resistance decade box. The output in volts is numerically equal to the
resistance in kilohms. The 0.1 µF capacitor acts as a noise rejection filter.

put is an exact copy of the cell voltage, but currents
of several milliamperes can now be drawn from the output without damage to the cell. Gain can also be built
in, as can be seen in the circuit of Figure 4-16b.
    A highly precise voltage source can be constructed
using combinations of followers and inverters with
gain, as illustrated in Figure 4-17. If the resistors
are of high quality, a precision of 0.1% is not difficult to attain. The Thevenin equivalent of the whole
circuit is a programmable voltage source with a very
low resistance in series.

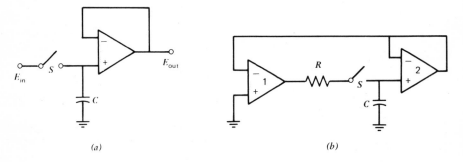

Figure 4-18. Sample-and-hold circuits. (a) One-amplifier version. (b) Improved version.

Another interesting application of the follower
is the *sample-and-hold* circuit shown in Figure 4-18a.
When the switch is closed, the voltage of the capacitor
tracks E_{in}. When the switch is opened, the capacitor
simply retains (holds) its charge for a considerable
period of time. A low-leakage capacitor must be used.

The rate at which the capacitor is charged depends on the maximum current I_{max} that can be drawn
from the source: $dE/dt = I_{max}/C$. To optimize this
parameter, a two-amplifier circuit can be used as
shown in Figure 4-18b, where the capacitor is charged
by the saturation voltage of amplifier 1, rather than
by E_{in} itself. Upon reaching equality with E_{in} the
amplifier is removed from saturation. Complete sample-
and-hold units are available in integrated circuit
form.

Another special purpose integrated circuit is the
multichannel programmable amplifier, for instance the
HA-2400 (Harris Semiconductor), shown schematically in

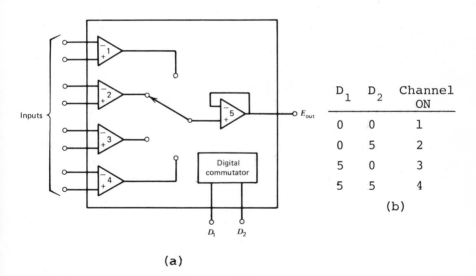

D_1	D_2	Channel ON
0	0	1
0	5	2
5	0	3
5	5	4

(b)

(a)

Figure 4-19. (a) Functional diagram of the HA-2400 programmable
op amp. (b) Selection code; voltages of zero or five volts applied to the D inputs will activate the designated channel.

Figure 4-19a. This consists of an array of four input
amplifiers, any one of which can be connected by a
fast internal switch to an output buffer amplifier.
The choice of channel is controlled by a built-in selector (*decoder* in digital parlance). Application of
a 5-V control signal to one, both, or neither of the
inputs D_1 and D_2 determines which input channel operates. Feedback must be provided from the output to
each input amplifier separately.

This module can be used in a programmable gain
mode, by simply connecting different resistors to give
R_f/R_{in} ratios of, say, 1, 10, 100, and 1000, selectable through the D inputs. It can also be used in a
continuously sequencing mode of operation (*multiplexing*), for example, to display four functions at the
same time on an oscilloscope. Because of the persist-
ence of the oscilloscope screen, the four consecutive
traces appear to be simultaneous.

NONLINEAR CIRCUITS

The operations discussed so far—addition, subtraction, differentiation, and integration—belong to
the category of *linear* operations. Linearity is defined by two properties:

(1) Two input signals E_1 and E_2 which, when applied individually, generate outputs E_3 and E_4, will
produce, when applied together, an output $E_3 + E_4$.

(2) An input signal E_1, which produces an output
E_3, when increased to kE_1, generates an output increased to kE_3.

For example, $E_{out} = E_{in}^2$ is a nonlinear relation,
since doubling E_{in} does not merely double E_{out}. In
contrast, it can easily be proved that integration is
a linear operation.

A very simple nonlinear circuit is the comparator, previously discussed and presented again in Figure 4-20. The output will be at either the positive
or negative saturation value, except for inputs in a

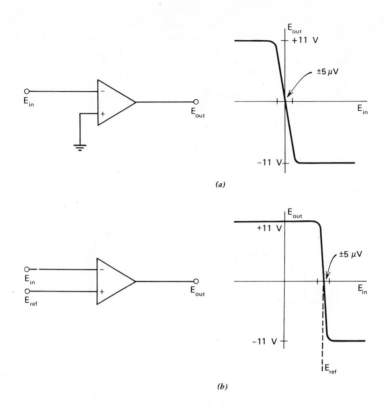

(a)

(b)

Figure 4-20. Comparators: (*a*) referred to ground, and (*b*) referred to E_{ref}. Amplifiers optimized for comparator use have very
fast transition times so that they can pass through the linear
region in considerably less than 1 μsec.

very small region (typically a few microvolts) where
the output will be intermediate between the two saturation limits. In this small region the behavior is
linear.

In Figure 4-20*b* is shown how the use of a differential mode permits the comparison of *two* voltages.

The comparator has numerous applications in logic
circuits. The abrupt switching between positive and
negative saturation can be used to trigger subsequent
digital devices. As an example, consider the case
where the time for the voltage from a transducer to

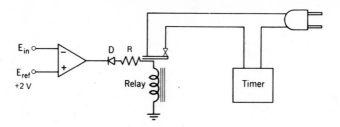

Figure 4-21. A timing circuit triggered by a comparator. The resistor R is needed to protect the amplifier from the so-called
inductive kick of the relay.

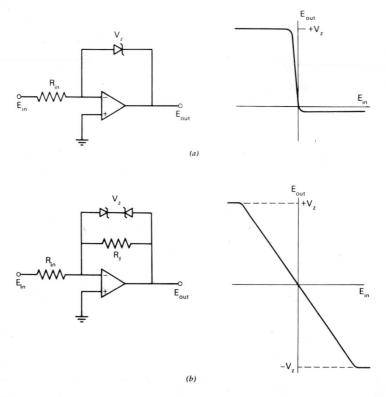

Figure 4-22. (a) Comparator with clamp limiting. The voltage
excursions are limited by V_z and by the forward voltage drop of
the diode, about −0.5 V. (b) Clamped summer with voltage limiting for both polarities; in the midregion the slope is given by
$-R_f/R_{in}$.

reach, say 2 V, must be precisely measured. The circuit in Figure 4-21 can be used for this purpose. As long as the value of E_{in} is less than E_{ref}, the output of the amplifier will be +11 V. Since the diode is now reverse biased, the relay remains in its normal (closed) condition. When the input becomes larger than the reference, the output switches to -11 V, the relay opens, and the timer stops.

Comparator applications using general purpose op amps may pose problems if the amplifier requires appreciable time to recover from saturation. In such cases it is preferable to limit the excursions of the output by means of a zener diode as feedback. Whenever the diode is conducting it provides feedback current. This limits the voltage excursions to the region where the diode does not conduct, that is, between zero and V_Z. This is illustrated in Figure 4-22a. If two identical zeners are used back-to-back, as in (b), the permitted voltages can range between $-V_Z$ and $+V_Z$. The limiting (clamping) action is not restricted to comparators, but can be used for summers as well. It is usually unsuited for integrators because of leakage in the diodes.

Another nonlinear circuit using diodes is the *precision rectifier* (Figure 4-23a). Note that the output is not taken directly from the amplifier, and might require a follower to avoid loading. When the input is negative, D_2 is forward biased and D_1 reverse biased. The equality of input and feedback currents requires that

$$\frac{E_{in}}{R_1} = -\frac{E_{out}}{R_2} \tag{4-24}$$

and since $R_1 = R_2$, it follows that $E_{out} = -E_{in}$. Note that the forward voltage drop of D_2 is immaterial.

This explains why this circuit can be called a *perfect rectifier*. The feedback path through D_1 becomes effective for positive inputs and takes over the feedback function exactly at zero volts, guaranteeing a sharp cut-off. Whereas a diode alone requires about 1 V for acceptable rectification, the perfect rectifier operates with as little as 10 mV.

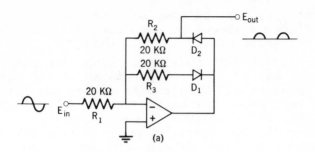

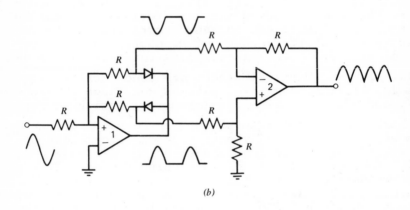

Figure 4-23. (a) Precision rectifier, (b) absolute-value circuit derived from (a), sometimes called a full-wave precision rectifier. In (b) the components associated with each amplifier should be carefully matched.

   The rectifiers we have been considering discard half of the available information contained in the signal. To utilize both halves, the circuit of Figure 4-23b can be used. This generates the *absolute value* of the input, by adding together the positive half and the inverted negative half. If a low-pass filter, such that $1/2\pi RC$ is less than the frequency of the signal, is inserted at the output, the result is a DC voltage proportional to the average value of the rectified AC. It can be made to drive an indicating meter, as in Figure 4-24.

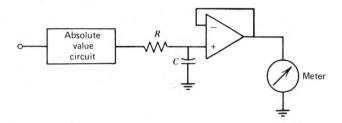

Figure 4-24. An electronic voltmeter using an absolute-value circuit. It responds to DC as well as to AC, with excellent response down to a few millivolts.

LOGARITHMIC FUNCTION GENERATION

The current-voltage relation for small signal diodes at room temperature is given by

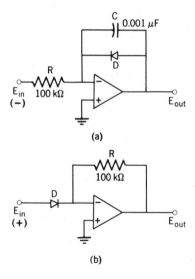

Figure 4-25. (a) Logarithmic converter, and (b) exponential (antilogarithmic) converter. The diode normally used is a silicon transistor with base and collector tied together ("transdiode"). Suitable types are 2N697, 2N1132, 2N2218 and 2N3900A.

$$E = B \log I + C \qquad (4\text{-}25)$$

where B and C are constants depending on the particular type of diode. This equation is obeyed fairly well by all diodes, and very closely by some specially selected types. The relationship suggests the use of diodes in generating both logarithmic and exponential functions.

An example is given in Figure 4-25a. Noting that $I_{in} = -E_{in}/R$, one can write

$$I_f = \frac{E_{in}}{R} \qquad (4\text{-}26)$$

and using Eq. (4-26), one obtains

$$E_{out} = B \log I_f + C = B \log \frac{E_{in}}{R} + C = B \log E_{in} + D \qquad (4\text{-}27)$$

where D includes all constants. This equation is obey-

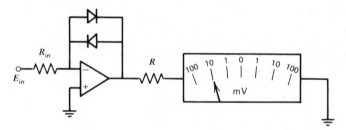

Figure 4-26.　Example of a logarithmic meter.

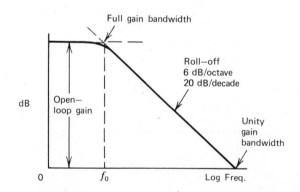

Figure 4-27.　Bandwidth definitions in an operational amplifier.

ed for a wide range of input voltages (typically 5 or 6 orders of magnitude). Observe that only one polarity of input signal is accepted. If the diode is connected at the input as shown in Figure 4-25b, the inverse function, the exponential or antilogarithm, is obtained.

The availability of log conversion modules permits the implementation of a logarithmic meter for monitoring signals of large dynamic range, as illustrated in Figure 4-26. This is useful as a null detector for DC bridges.

OP AMP PARAMETERS

Operational amplifiers can be described by a number of parameters, some of which have been mentioned previously, some not. The important parameters are listed in Table 4-1 for reference, with the desirable order of magnitude for an ideal amplifier, also with ranges of values normally found in commercial units. These parameters can be defined as follows.

The *open-loop gain*, also called *large-signal voltage gain*, is the ratio of E_{out} to the voltage difference between the two inputs. This is a function of frequency (Figure 4-27).

The *differential input impedance* is the effective internal resistance between the two inputs (Figure 4-28).

The *common-mode input impedance* is the internal resistance from either input to ground.

The *output impedance* is the Thevenin impedance, the amplifier being considered a source of voltage driven by the input (Figure 4-28). This quantity becomes much smaller in the presence of negative feedback.

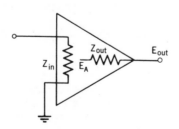

Figure 4-28. Input and output impedances of an amplifier.

Table 4-1

Operational Amplifier Parameters

Parameter	Ideal	Typical
Open-loop gain (A or A_{vo})	∞	20–1000 V/mV
Input impedance (Z_{in})	∞	0.1–100 MΩ
For FET input amplifiers		10^4–10^6 MΩ
Common-mode input imp. (Z_{in-CM})	∞	1–10^6 MΩ
Output impedance (Z_{out})	0	10–100 Ω
Unity-gain bandwidth (BW)	∞	1–1000 MHz
Offset voltage (V_{os})	0	0.1–10 mV
Temp. coefficient of V_{os} (TCV$_{os}$)	0	5–500 μV/K
Input bias current (I_B)	0	0.1 nA–10 μA
Input offset current (I_{os})	0	0.1 nA–10 μA
Rise time (RT)	0	10 ns–10 μs
Slew rate (SR)	∞	0.1–100 V/μs
Settling time (to 0.1 %) (t_s)	0	50 ns–50 μs
Common-mode rejection ratio (CMRR)	∞	60–120 dB
Power supply rejection ratio (PSRR)	∞	50–100 dB
Price (for small quantities)	0	0.5–20 $

    The *unity-gain bandwidth* is the range of frequencies from DC to that point where the open-loop gain becomes unity (zero dB) (Figure 4-27).

    The *offset voltage* is an unwanted internal potential between the two inputs, seen at the output as

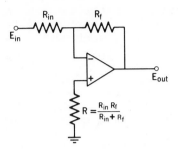

Figure 4-29.   A method for minimizing bias effects.

multiplied by the circuit gain.   It can be minimized
(nulled) by a trimming potentiometer used to balance
the input circuit of the amplifier.   The offset voltage is usually temperature dependent, and so nulling
at one temperature may not hold at another.
     The *input bias current* is the average of the re-
sidual currents flowing into the two inputs under normal operating conditions and with no signal present.
A certain minute current must always flow into the inputs to carry information.   In Figure 4-29 is shown a

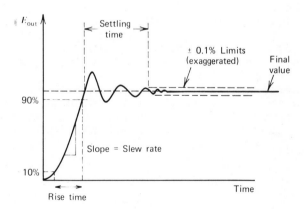

Figure 4-30.   Illustration of the dynamic behavior of an operational amplifier connected as a summer.   The curve represents the
output response to an input step.   The oscillations around the
final value are exaggerated for clarity.

method of correcting for the effects of bias current by means of a resistor R, connected from the noninverting input to ground, such that $R = (R_{in}R_f)/(R_{in} + R_f)$, the parallel combination of R_{in} and R_f. FET-input amplifiers do not need this correction.

The *input offset current* is the difference between the two input bias currents.

The *rise time* is the minimum time required for the response to a voltage step to go from 10% to 90% of the resulting output step (Figure 4-30).

The *slew rate* is the average slope of the step-response curve in going from 10% to 90%. This represents the fastest rate at which the amplifier can respond (Figure 4-30).

The *settling time* is the time required for the response to a step signal to proceed to the point where the error is permanently within specified limits, such as ±0.1% (Figure 4-30).

The *common-mode rejection ratio* occurs when a voltage E_{CM} is applied equally to both inputs (common mode). Ideally the output will not change, since the amplifier responds only to the difference between the two inputs. However, it usually does change by a small amount. This can be represented by $E_{CM}/CMRR$, where CMRR is the common-mode rejection ratio (Figure 4-31).

The *power-supply rejection ratio* is the ratio of a change in power supply voltage to the resulting offset voltage. This is especially important in battery-operated systems, where the supply voltage is apt to change with time.

Commercial op amps are made in many types with variations in their parameters to optimize them for specific services. General purpose units should be selected unless the requirements are particularly exact-

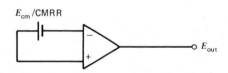

Figure 4-31. Model for common-mode voltage. The net effect on the output is indistinguishable from an offset of $E_{cm}/CMRR$.

ing. The majority of amplifiers are "bipolar," whereas others are "FET" types. (These designations refer
to the variety of transistors in their input circuits.)
FET-input amplifiers have very high input impedance
and correspondingly low input bias currents. Bipolars,
on the other hand, tend to have lower voltage offset
and lower price. Special amplifiers are available
optimized for use as voltage followers. Others have
the response speed maximized. Comparators, particularly, are often required to have a very fast response,
and special units are to be preferred to general purpose op amps for this service.

In Table 4-2, the important specifications are
given for selected commercial amplifiers. Type 741 is
the most widely used op amp at present. Characteristics of the basic model are given in Column 1. It is
manufactured by numerous firms with various improve-
ments. The model 741K of Analog Devices (Column 2)
has a reduced voltage offset which usually allows one
to dispense with the nulling potentiometer, an inconvenient and expensive item. The FET amplifier 3521J
(Burr-Brown) combines low voltage offset and outstand-
ingly high input impedance. The OP-05CP (Precision
Monolithics) is an unusually high-performance amplifier
for low-frequency applications. It does not normally
require offset nulling, its noise and drift are very
low, and its high CMRR makes it useful in differential
applications. The µA715 (Fairchild) is a fast amplifier with moderately high offsets. It is useful for
comparator and fast sample-and-hold circuits because
of its short rise time of 200 nsec.

All the amplifiers listed in Table 4-2 except the
µA715 are limited in their AC response to about 10 kHz,
because of the gain roll-off in the Bode plot. Beyond
that, up to the unity gain point, circuits will deviate increasingly from ideality. Thus Eq. (4-4) will
no longer reduce to Eq. (4-5), distortion will be introduced, and the maximum amplitude will be limited.
The µA715, however, can be used as far as 200 kHz without difficulty.

We recommend that a person or laboratory contem-
plating instrument work standardize on a few types of
amplifiers, selected to have identical pin connections
and thus to be directly interchangeable. A good selection that will cover a large proportion of applications would be the 741K, 3521, and OP-05CP, used without nulling circuits.

TABLE 4-2

Comparative Characteristics of Selected Operational Amplifiers   (Typical Values)

Model	741C	741K	3521J	OP-05CP	μA715	Units
Manfacturer[a]	Many	AD	BB	PMI	F	
Open-loop gain	200	200	50	400	30	V/mV
Input imp.	2	2	10^5	33	1	MΩ
Output imp.	75	75	1000	60	75	Ω
Offset volts	2	0.5	0.25	0.3	2	mV
Offset TC	—	0.006	0.005	0.002	—	mV/°C
Bias current	80	30	0.02	1.8	400	nA
Offset curr.	20	2	0.002	1.8	70	nA
CMRR	90	100	90	120	92	dB
PSRR	90	—	—	104	83	dB
Bandwidth	1	1	1.5	0.6	65	MHz
Max output curr.	25	25	10	50	—	mA
Slew rate	0.5	0.5	0.8	0.17	100	V/μs
Price (approx.)	1	4	20	4	—	$

[a]AD: Analog Devices, Norwood, MA 02062
BB: Burr-Brown Research Co., Tucson, AZ 85706
PMI: Precision Monolithics, Inc., Santa Clara, CA 95050
F: Fairchild, Palo Alto, CA 94304

ERRORS

It should be pointed out that an op amp, per se
(without feedback), may deviate from linearity by several percent. This and other kinds of errors are diminished when negative feedback is applied. Figure
4-32 shows the Bode plot for a typical op amp with an
open-loop gain of 100 dB. The lower curve is obtained
when the loop is closed at 60 dB, as in a summer with
a gain of 1000. Note that the horizontal part of the
curve now extends approximately 10^3 Hz, whereas without feedback, the curve falls off at 10 Hz. The gain
difference (40 dB, called the *loop gain* or the *gain
margin*) is the effective feedback.

The gain margin, representing the excess gain of
the amplifier proper over the circuit gain, is the

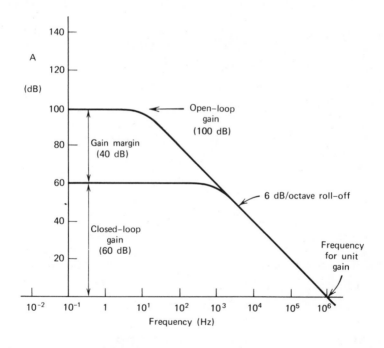

Figure 4-32. The relation between open-loop and closed-loop gains.
Observe that a limitation in the circuit gain produces an expansion of the horizontal portion of the curve.

actual factor that corrects errors due to nonlinearity.
It can be shown that most nonlinearities and distortions inherent in the electronics of the amplifier are
diminished in proportion to the gain margin. In figure
4-32, if the amplifier has a 5% inherent distortion
before feedback, the circuit with feedback will have
5%/100 = 0.05% distortion. This applies from DC to
approximately 10 Hz, but the distortion worsens thereafter until, at 1 kHz, there is no more gain margin
available. The frequency range over which the amplifier is useful will extend to perhaps 100 Hz, depending upon the permissible distortion. Note that this
calculation applies to a circuit with R_f/R_{in} = 1000.
The same amplifier at unity gain is usable to perhaps
10 kHz.

Under nonlinear conditions two main types of distortion can appear: (1) *Harmonic distortion*—this
consists of generation of multiples of the original
frequency (harmonics). (2) *Intermodulation distortion*
—this appears in addition to the above when two frequencies f_1 and f_2 are present together. It consists
of the sum and difference frequencies, $f_1 \pm f_2$. It
can be very annoying, since it generates specific spurious frequencies that can be mistaken for signals. DC
errors due to nonlinearity are usually not observed
because of the gain margin. Distortions of square
waves and pulses can be predicted from the step response shown in Figure 4-30.

INTEGRATOR ERRORS

It is essential for the effective use of integrators to compute the extent of different offset and
bias errors (Figure 4-33). For high-quality capacitors, as should be used in integrators, the leakage
resistance R_{leak} is about 10^{11} $\Omega/\mu F$. Let us assume
that a rather good quality amplifier is employed, with
E_{os} = 0.05 mV, and the bias current I_B = 0.05 nA. The
operation of the circuit is such that

$$I_{in} = -C \frac{dE_{out}}{dt} \qquad (4-28)$$

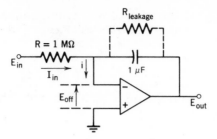

Figure 4-33. An integrator circuit showing the sources of error.

In the presence of the error sources indicated, one must add to I_{in} the error currents

$$I_{error} = I_B + \frac{E_{os}}{R_{in}} + \frac{E_{out}}{R_{leak}} \qquad (4\text{-}29)$$

In the example, for $E_{out} = 10$ V, the error current* is

$$I_{error} = 0.05 \times 10^{-9} + \frac{0.05 \times 10^{-3}}{1 \times 10^6} + \frac{10}{10^{11}}$$

$$= 2 \times 10^{-10} \text{ A} = 0.2 \text{ nA} \qquad (4\text{-}30)$$

The error current will charge the capacitor at the rate of $(2 \times 10^{-10})/(1 \times 10^{-6}) = 0.2$ mV/sec. In practice this means that for a maximum error of one percent, the slowest permissible rate of integration is 20 mV/sec, or about 1 V/min. This indicates that integrations over long periods of time (hours) are usually not feasible with op amps.

PROBLEMS

4-A. Design a circuit to perform the following data-processing task:

$$E_{out} = 5E_1 + 7E_2 - 14E_3$$

*This assumes the worst case, when the errors add rather than partially cancelling each other.

4-B. Write the algebraic equation describing the operation of the circuit of Figure 4-34.

4-C. Describe the functioning of the circuit of Figure 4-35.

4-D. Write the operating equation for the circuit of Figure 4-36.

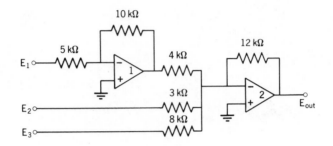

Figure 4-34. See Problem 4-B.

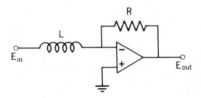

Figure 4-35. See Problem 4-C.

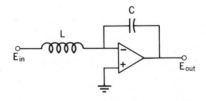

Figure 4-36. See Problem 4-D.

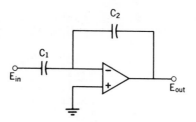

Figure 4-37.   See Problem 4-E.

4-E.   Describe mathematically the circuit of Figure
       4-37.

4-F.   Show that both differentiation and integration
       are linear operations.

4-G.   It is desired to compute a voltage E as a func-
       tion of two variable quantities P and Q, and a
       constant K, according to the equation:

$$E = K + 0.059 \log \frac{P}{Q}$$

       Design an op amp circuit to carry out this com-
       putation.

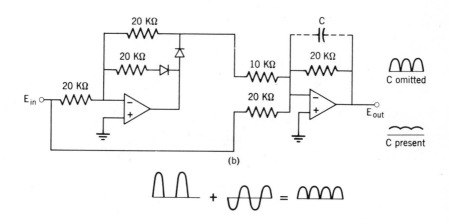

Figure 4-38.   See Problem 4-H.

4-H.   Figure 4-38 shows an alternative circuit for a
       perfect rectifier. Explain how it works.

4-I.   Make the error calculations for the difference
       integrator of Figure 4-12a.

4-J.   Design the input and feedback circuits for the
       HA-2400 amplifier of Figure 4-19 to give pro-
       grammable gains of 1, 3, 10, and 30.

*   *   *

4-1.   Design an operational amplifier circuit to per-
       form the operation:

$$-E_{out} = 5E_1 + 3E_2 + E_3$$

4-2.   Write the equation for the circuit of Figure 4-
       39.

4-3.   Design a circuit to give E_{out} according to the
       following equation, where E_{in} is in volts and I_{in}
       in amperes:

$$E_{out} = -8E_{in} - 10^6 I_{in}$$

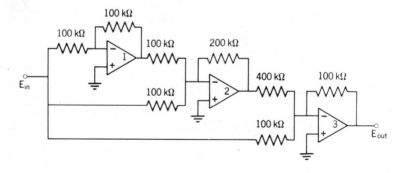

Figure 4-39.   See Problem 4-2.

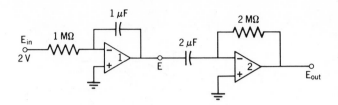

Figure 4-40. See Problem 4-6.

4-4. Write the relations between input and output as functions of the parameter k for the circuits of Figure 4-5b, c and d.

4-5. What voltage will be present at the output of an integrator 5 min after start of integration if the input resistor is 600 kΩ, the feedback capacitor is 0.5 µF, and the input voltage is 1 mV?

4-6. Evaluate E_{out} in the circuit of Figure 4-40 for the component values marked.

4-7. Describe and sketch the input-output relations

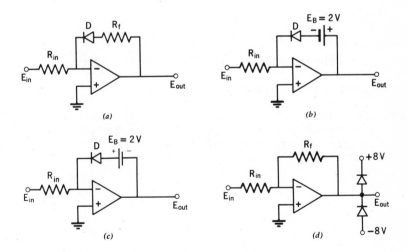

Figure 4-41. See Problem 4-7.

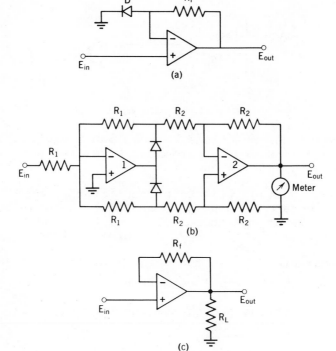

Figure 4-42.   See Problem 4-8.

for the circuits of Figure 4-41.   Assume that
saturation occurs at ±10 V;   neglect the forward
drop of the diodes.

4-8.   Describe (and sketch where appropriate) the
input-output relations for the circuits given in
Figure 4-42, with the same assumptions as in
Problem 4-7.

V

ANALOG INSTRUMENTATION

Operational amplifiers can be used with feedback around a group of several units, in addition to the individual feedback loops. Such multiple feedback is used mostly for generating signals that are functions of time, sometimes very complex functions.

To understand how this is done, consider a pair of integrators followed by an inverter (Figure 5-1a). The output is calculated by combining the individual expressions shown in the figure which gives

$$x = -\frac{1}{R_1 C_1 R_2 C_2} \int\int E_{in} \, dt^2 \qquad (5-1)$$

to which must be added the integration constants.

Equation (5-1) is equivalent to the relation

$$E_{in} = -k \frac{d^2 x}{dt^2} \qquad (5-2)$$

where $k = R_1 C_1 R_2 C_2$. If a loop is now closed in the circuit between the points marked P and Q, a powerful restriction is introduced into Eq. (5-2). Of all values of x, only those are now possible for which $x = E_{in}$, or

$$x = -k \frac{d^2 x}{dt^2} \tag{5-3}$$

We thus come to the conclusion that by connecting P and Q we have limited the behavior of the circuit to obeying the differential equation (5-3). The output of amplifier 2 describes the solution of the equation, multiplied by a constant $1/k$. The solution of this particular differential equation is a sine wave:

$$x = A \sin (\omega t + \phi) \tag{5-4}$$

in which one can show that $\omega = 1/\sqrt{k} = 1/(\sqrt{R_1 C_1 R_2 C_2})$ the circuit of Figure 5-1b can be used to generate a sine wave, with frequency determined by the RC time constants.

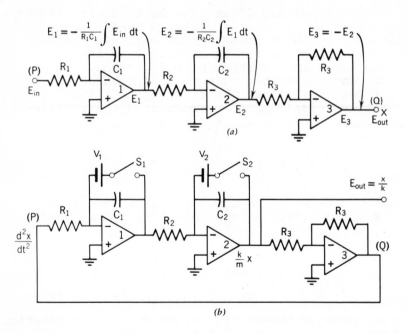

Figure 5-1. (a) A circuit for double integration with sign inversion. (b) An oscillator made by closing a loop from Q to P. The initial conditions should be $V_1 = 0$ for integrator 1 and V_2 equal to the desired amplitude, for integrator 2.

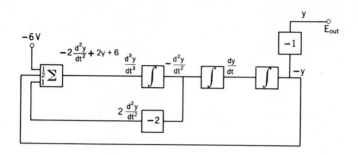

Figure 5-2. Solution of a third-order differential equation. The
boxes represent integrators and inverters of the types previously
discussed. The box marked Σ contains a summer with gains of 1,
2, and 10, respectively, at its three inputs.

One factor that requires discussion is the presence of integrating constants. They have an electronic significance, the counterparts of their mathematical meanings; they represent the initial values
of the capacitor voltages, V_1 and V_2 in the figure.
For every combination of V_1 and V_2, there corresponds
a particular pair of constants A and ϕ.
This sine-wave generator is not particularly
stable in amplitude, and other designs are usually
preferred, but the method is very convenient to generate functions that are solutions of more complex differential equations. As an example, Figure 5-2 shows
how one may solve the equation.

$$\frac{d^3y}{dt^3} = -2\frac{d^2y}{dt^2} + 2y + 6 \qquad (5-5)$$

A direct connection as before ensures that the equation is obeyed. A plot of the function $y(t)$ can be
obtained by connecting a recorder at E_{out}.
Linear differential equations of any order can be
solved by this general method. To solve nonlinear
equations, special components are necessary, the most
important of them being the multiplier, divider, and
function generator (Figure 5-3).
The *multiplier* is a device generating the product
of two variables X and Y, such that the output is

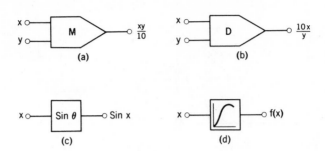

Figure 5-3. Nonlinear elements in analog computation: (a) multiplier, (b) divider, (c) sine-function, and (d) an arbitrary function generator.

$$E_{out} = \frac{XY}{10} \tag{5-6}$$

The division by 10 is necessary to utilize fully the dynamic range of the amplifiers. Thus if both X and Y are 10 V (their upper permissible limit), the output will be 10 V, not 100 V. The *divider* can perform the inverse function, actually giving $10Y/X$. In some cases division and multiplication can be implemented by the same component.

Function generators can be designed to produce a specified mathematical response (sine, logarithm, square, etc.). Some function generators are flexible, permitting the operator to program arbitrary functions.

ANALOG COMPUTERS

An analog computer (Figure 5-4) is a collection of operational amplifiers wired to a common panel, so that a large number of circuits can be synthesized by interconnecting (patching) different points on the panel. In addition to the amplifiers, there are usually a number of potentiometers (to provide constant coefficients), nonlinear components, and read-out devices. More advanced computers, such as that in the figure, contain also digital logic elements.

In addition to their utility as summers, the am-

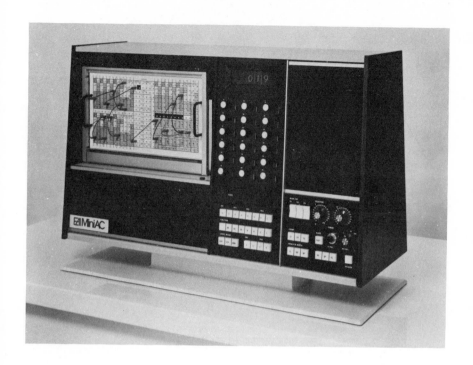

Figure 5-4. A 27-amplifier analog computer containing also digital logic capability. The EAI "MiniAC" (Photo courtesy of Electronic Associates, Inc., West Long Branch, NJ.)

plifiers can be connected as integrators, in which case a central relay system provides synchronized reset, integrate, and hold modes. In the reset mode, the amplifiers are brought to the appropriate initial conditions, which must be specified for each integrator.

Analog computers are available in various sizes from elaborate systems with hundreds of amplifiers down to small units with 24 amplifiers. The latter can be used to great advantage for testing out op-amp circuits in prototype instruments, prior to actual construction. A second application of analog computers is in solving complicated differential equations, even those for which there is no mathematical solution. Finally, analog computers are useful in *simulation*, the synthesis of circuits that follow the same behavior as real systems.

For the purpose of illustrating the use of an ana-

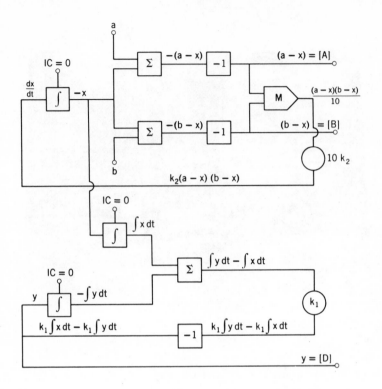

Figure 5-5. Solution of the differential equations for the kinetic problem described in the text. Note that $[C] = (x - y)$; this
could be obtained from the computer with one additional summer.
The letters IC denote initial conditions at the several integrators. Circles represent attenuating potentiometers. Brackets
denote concentrations.

log computer to simulate a physical system, consider
two consecutive chemical reactions

$$A \ + \ B \ \xrightarrow{\ (1)\ } \ C \ \xrightarrow{\ (2)\ } \ D \qquad (5\text{-}7)$$

Let the initial conditions be $A(0) = a$, $B(0) = b$,
$C(0) = 0$, and $D(0) = 0$. It is important to know the
exact initial conditions, since the analog computer
does not produce the *general* solution, but only one
particular graphic solution characterized by the given
initial conditions.

The equations to be solved are

$$\frac{dx}{dt} = k_1 (a - x)(b - x) \qquad\qquad (5\text{-}8)$$

and

$$\frac{dy}{dt} = k_2 (x - y) \qquad\qquad (5\text{-}9)$$

where x is the amount of A or B that has reacted at time t, and y is the quantity of D formed. The proportionality constants, k_1 and k_2, refer to reactions 1 and 2, respectively.

It is advantageous to rewrite Eq. (5-9) as

$$y = k_2 \int x \, dt - k_2 \int y \, dt \qquad\qquad (5\text{-}10)$$

where the integration constants $x(0)$ and $y(0)$ are both zero.

In the computer setup of Figure 5-5, the upper section solves for the value of x (the extent of the first reaction) independently of y. This is possible because we have assumed no reverse reaction. In the lower loop, y is computed in terms of the quantity x obtained from the upper section.

In practice the circuit can be made considerably simpler by combining functions, for example by using summing integrators.

The solutions can be plotted directly if k_1 and k_2 are known. Otherwise a succession of runs can be made (repetitive mode) while the operator changes the values of the k's by adjusting the two potentiometers, until the curve matches the experimentally obtained data. This type of curve fitting is very useful when the equations are too complex for a mathematical solution.

When using an analog computer, it is important to choose the proper values for the coefficients relating the voltages to physical quantities. For example, 1 V might equal 9 moles/liter. This operation, called *scaling*, determines the voltage excursions that each amplifier will make during the computation. One must make certain that no amplifier will be asked to have an output larger than its saturation level at any time, or else errors will exist. Scaling is often the most troublesome stage in analog computation.

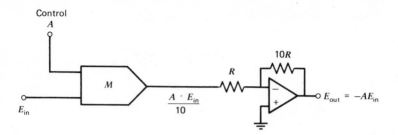

Figure 5-6. A multiplier used as a variable-gain amplifier. The op amp is needed in order to make the gain numerically equal to the value of A in volts.

MORE ABOUT MULTIPLIERS

In addition to their use in function generation, multipliers have found many other interesting applications. In the past multipliers were expensive 20-pounders of rather slow and temperamental operation. There was no reason to think of them as general purpose components. In contrast, a modern multiplier is a single integrated circuit, moderately priced, and capable of operating at the 0.5% or better error level. It also provides divider operation merely by changing external connections.

A useful application of a multiplier is in voltage-controlled amplification (Figure 5-6). The gain is adjusted by a voltage at A, and can be changed rapidly to a new value simply by adjusting the voltage control. This can be useful, for example, in autoranging meters, where the sensitivity of an instrument adjusts itself automatically, depending on the input level.

If the control voltage at A is a sine wave, the effect is amplitude modulation (Figure 5-7). The resulting curve has an amplitude of E_{in} modulated by the control wave. Sine-wave modulation, as discussed previously, generates the sum and difference of the two frequencies f_1 and f_2 involved (that is, $f_1 \pm f_2$, Figure 5-8). This is very useful for generating sine waves of new frequencies, and is especially interesting, as shown in Figure 5-8b, for doubling the frequency (if $f_1 = f_2$). The double frequency (second harmonic) tracks any shifts in the original frequency.

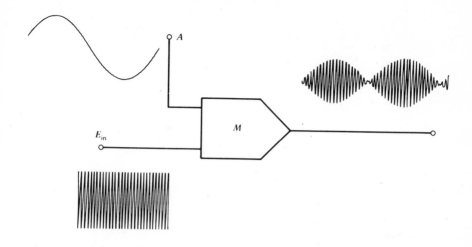

Figure 5-7. Modulation by means of a multiplier. A so-called
four-quadrant multiplier must be used to accomodate both positive
and negative polarities.

A variety of waveshapes can be used for modulation in addition to sine waves, the most important being the square wave. If a symmetrical square wave of,
say, +1 to -1 V is used, the result is a periodic sign
inversion, (*chopping*). If the chopping is carefully

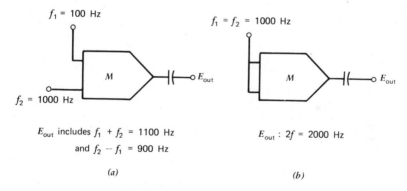

Figure 5-8. Multipliers used to synthesize new frequencies. (*a*)
Sum and difference, (*b*) second harmonic. The capacitor serves to
remove a DC component generated by this process of multiplication.

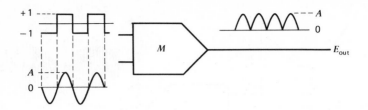

Figure 5-9.   Example of synchronous detection.

made to coincide with the zero-crossing of the input
sinewave, the result is a type of full-wave rectification called *synchronous detection* (Figure 5-9). The
use of synchronous detection in lock-in amplifiers is
discussed later in this chapter.

ACTIVE FILTERS

        In this section we discuss a number of devices
that serve to modify signals to render their information content more easily utilized. As black boxes,
they accept input signals and produce outputs related
thereto according to some set plan or program. A good
example is the active filter.
        The filters described earlier are constructed en-
tirely of passive components, and this restricts their
usefulness. The same R and C components can be used
in frequency-selective circuits with op amps to achieve
similar and often superior results. These are called
active filters. Figure 5-10 shows (at a and b) low-
pass and high-pass versions that provide adjustable
gain (R_2/R_1), not possible with passive filters. Figure 5-10c shows a band-pass filter that combines the
actions of the first two to give attenuation at both
high and low frequencies. This cannot be accomplished
readily with a passive RC filter network.
        The active filters in Figure 5-11 give twice the
attenuation slope, 40 dB/decade, rather than 20. The
response of the band-pass unit (Figure 5-11c) can be
tailored for various slopes. These filters are referred to as *second order*.
        There are many possible variants on second- and
higher order filters. An example of an easily constructed low-pass filter with f_0 = 1 Hz, very effective

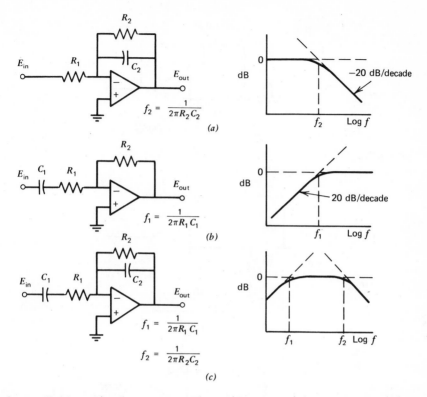

Figure 5-10. Single-stage active filters. (a) Low-pass, (b) high-pass, and (c) band-pass.

in removing noise from a DC signal, is given in Figure 5-12. This filter gives no appreciable attenuation from DC to 0.5 Hz, yet more than 90% attenuation at 2 Hz and essentially complete beyond about 10 Hz. Any other value of f_0 can be obtained merely by multiplying all resistance values by the new $1/f_0$.

The Twin-Tee notch filter described in Chapter III can be placed in the feedback loop of an op amp (Figure 5-13) to give the inverse of a notch, a sharply tuned band-pass filter, usually thought of as a *tuned amplifier*. The Twin-Tee network provides a low impedance for all frequencies other than f_0, and a high impedance at that frequency. Hence the closed-loop gain of the amplifier is high at f_0 and low elsewhere.

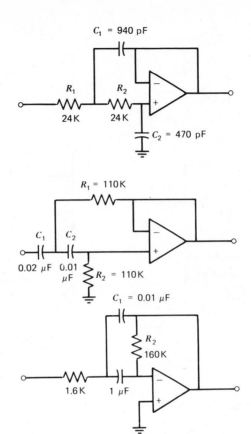

Figure 5-11. Second-order active filters. (*a*) Low-pass, (*b*)
high-pass, and (*c*) band-pass. This band-pass filter gives a sharp
peak at the center frequency, rather than the wide band seen in
Figure 5-10*c*. (From E. R. Hnatek, "Applications of Linear Integrated Circuits, " by permission of John Wiley & Sons)

The 22-MΩ resistor paralleling the Twin-Tee limits the
gain at resonance to avoid saturation.

MODULATION

We have seen that as one approaches DC, noise increases substantially, following a $1/f$ variation with
frequency. By means of modulation (mentioned earlier),

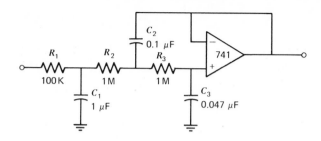

Figure 5-12. A second-order low-pass noise-rejection filter. Interchanging resistors and capacitors will give a high-pass filter.

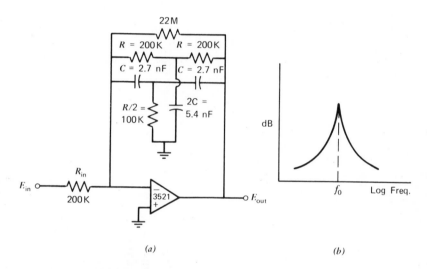

Figure 5-13. (a) A tuned amplifier using a Twin-Tee feedback filter. The values given correspond to a center frequency of f_0 = 300 Hz, determined from the formula $f_0 = 1/(2\pi RC)$. The resistors should be matched to ±0.1%, capacitors to ±1% or better. (b) Frequency response.

it is possible to transfer information from low frequency to a carrier of higher frequency which brings it into a relatively noise-free region. Wave forms corresponding to AM and FM are shown in Figure 5-14 and 5-15, respectively.

Modulation is widely used in communications systems (radio, TV, etc.) where specific carrier fre-

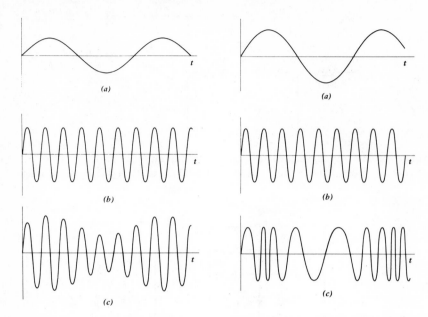

Figure 5-14. Example of amplitude modulation: (*a*) signal, (*b*) carrier, and (*c*) modulated wave.

Figure 5-15. Frequency modulation: (*a*) signal, (*b*) carrier, and (*c*) modulated wave.

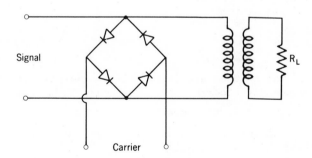

Figure 5-16. A diode modulator. A transformer is often used as shown for coupling the modulator to the load. Band-pass filters are usually added to reject the residual unmodulated signal and various harmonics. The diode ring is available as an integrated circuit.

quencies and bandwidths are officially assigned to
avoid interference. In measurement systems the frequency of the carrier can be selected to give the best
S/N ratio in a given situation. A relatively low carrier frequency (e.g., 1000 Hz) is often used. After
amplification, the information can be returned to the
original form by the process of demodulation.

There are numerous circuits for modulation. One
of them is shown in Figure 5-7. For higher frequencies
than a multiplier can handle, one can use the diode
modulator shown in Figure 5-16. In both cases, modulation amounts to generating product terms of the type
$E_1 E_2 \sin \omega_1 t \sin \omega_2 t$. The nonlinear current-voltage
curve of a diode can be expressed as a power series

$$I = k_1 E + k_2 E^2 + \ldots \qquad (5\text{-}11)$$

The first term describes the behavior of a linear device; for example, a resistor, where $k_1 = 1/R$ (Ohm's
law). The second term is responsible for modulation.
If two frequencies are present, ω_1 and ω_2, the total
current is given by

$$I = k_1 (E_1 \sin \omega_1 t + E_2 \sin \omega_2 t)$$

$$+ k_2 (E_1 \sin \omega_1 t + E_2 \sin \omega_2 t)^2 + \ldots \qquad (5\text{-}12)$$

which can be rewritten as

$$I = k_1 E_1 \sin \omega_1 t + k_1 E_2 \sin \omega_2 t + k_2 E_1^2 \sin^2 \omega_1 t +$$

$$k_2 E_2^2 \sin^2 \omega_2 t + 2 k_2 E_1 E_2 \sin \omega_1 t \sin \omega_2 t + \ldots \quad (5\text{-}13)$$

The first two terms represent unchanged initial frequencies. The following two are found (by means of the
identity $\sin^2 x = 1/2 - 1/2 \cos 2x$) to consist of a DC
current and of the second harmonics of the original
frequencies. The presence of a DC component is to be
expected in a rectifying system. The term containing
the product of sines represents the modulated wave.

It now becomes evident that a tuned filter is
necessary not only for noise rejection, but also to

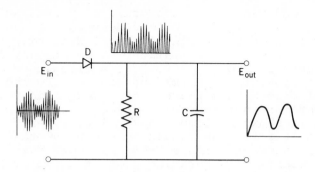

Figure 5-17. A diode detector. The diode rectifies the input
signal, which is then smoothed out by the capacitor to give the
envelope (the same shape as the original signal). The carrier
must be of much higher frequency than the signal, so that the *RC*-
filter attenuates the former but not the latter.

eliminate the various harmonic frequencies as well as
the original signal. The unmodulated carrier component is usually left in, although some techniques, including the multipliers, suppress it. In the circuit
of Figure 5-16, the transformer removes the DC component.

A modulated signal must be demodulated to extract
the original information. A simple demodulation circuit is the *diode detector* shown in Figure 5-17. The
diode, this time operating at higher voltage levels
than in the modulator, acts as a rectifier, producing
a train of half-waves. The RC filter smoothes the
series of pulses to a faithful replica of the original
signal. The effect is more precise the larger the
difference between the frequency of the signal and
that of the carrier. If, for example, the carrier is
1 MHz and the signal is 100 Hz, there are 10,000 pulses
in each cycle of the detected signal. Following the
filter, the 100 Hz signal will be essentially free of
residual ripple. For low to moderate frequencies, the
perfect rectifier makes an excellent demodulator.

Another type of modulation of importance in instrumentation is *pulse modulation*. The repetition rate
of the pulses can be considered analogous to the frequency of a sine wave. The train of pulses can be modulated either in amplitude, in duration, or in position, as shown in Figure 5-18.

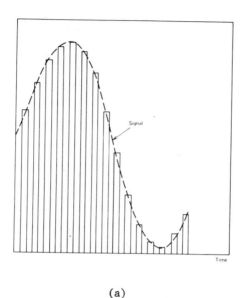

(a)

Figure 5-18a. Pulse amplitude modulation. All pulses are of the same duration.

Pulse amplitude modulation is used in the so-called *chopper-stabilized amplifiers* (Figure 5-19). Note that the complete system is designed as a single-input amplifier (the large triangle). It consists of two parallel lines: Amplifier 1 is driven by a chopped version of the input, and after demodulation feeds a very stable DC output into amplifier 2. A separate path, provided with a high-pass filter, brings the AC component of the signal directly into amplifier 2,

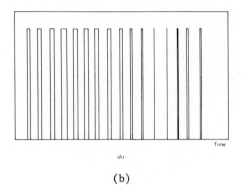

(b)

Figure 5-18*b*, Pulse width modulation. The pulses vary in duration.

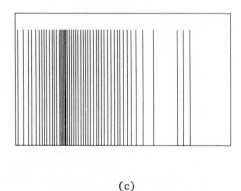

(c)

Figure 5-18*c*. Pulse position modulation. The pulses are of uniform duration but vary in spacing.

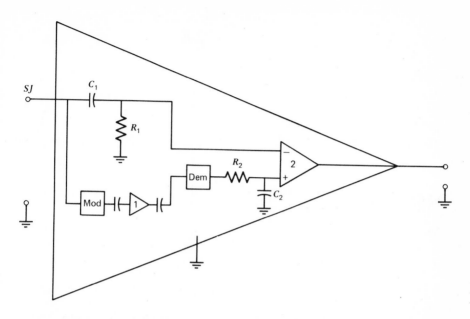

Figure 5-19. Chopper-stabilized amplifier. The sole input connection is the summing junction. Conventional input and feedback
circuitry is required.

where the two components are effectively summed together to give the final output. The overall combination
has a very high open-loop gain ($\sim10^7$), and an outstanding stability (~5 μV/year). On the other hand, this
type of amplifier is prone to slow recovery from saturation (perhaps several seconds).

LOCK-IN AMPLIFIERS

A special type of amplitude demodulation is found
in lock-in amplifiers, used extensively in measuring
weak AC signals. Lock-in amplifiers are based on the
synchronous detector mentioned earlier (Figure 5-9).
The lock-in amplifier requires, in addition to the
signal itself, a reference that provides frequency and
phase information for the synchronous detector. The
reference signal can be made quite strong, hence im-
mune from interference.

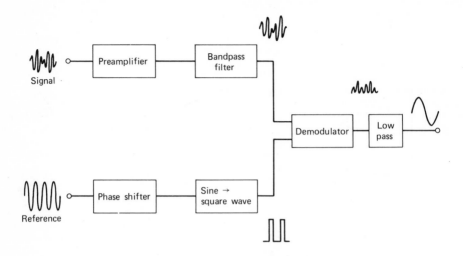

Figure 5-20. Block diagram of a typical lock-in amplifier.

A typical lock-in amplifier diagram is shown in
Figure 5-20. It consists of the following parts: (1)
A low-noise preamplifier, important to permit measuring
submicrovolt signals.   (2) A band-pass filter tuned to
the frequency of the carrier.   (3) A phase-shifter,
which advances or retards the phase of the reference to
allow for phase shifts or time delays in the system.

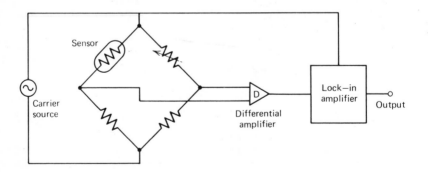

Figure 5-21. Lock-in bridge system. This circuit is very sensitive and responds to resistance changes of a few parts per million.
Some commercial lock-in amplifiers have a built-in carrier oscillator.

(4) A sine-to-square wave converter, which could be a simple comparator to sense the zero crossing of the wave. (5) The synchronous demodulator. (6) A low-pass filter to eliminate the carrier frequency.

Lock-in amplifiers are widely used for measuring low-level signals. The requirement that the signal be modulated on a carrier can be fulfilled simply by using the carrier as a stimulus to the system, as illustrated in Figure 5-21. The bridge, after manual balancing, will give an output proportional to any resistance change in the sensor, which could correspond to differences in temperature, light intensity, etc. It should be noted that one could use a diode detector or a perfect rectifier in place of the lock-in amplifier, except for very low-level signals. The perfect rectifier cuts off automatically when the signal becomes negative and cuts in automatically when it is positive, in exactly the same manner as does the synchronous rectifier. One can say that the demodulator is internally synchronized. The trouble begins when the noise exceeds the signal. The diode then cuts off whenever the *noise* voltage becomes negative, so that the noise rather than the signal is synchronously rectified. (The signal is actually attenuated.) The merit of the lock-in amplifier is that, regardless of how large the interference is, the detection always takes place with the proper phase. This is possible because, unlike the signal, the reference is at a high level compared with the noise.

SIGNAL AVERAGING

In many measuring systems it is possible to trigger a process repeatedly by a suitable excitation. For example, in a luminescence measurement a flash of light may be applied and the time dependence of the resulting response monitored. On repetition, the signal will be identical each time and superimposed upon itself. Many physiological experiments are conducted in this manner.

A set of response curves could be averaged manually point by point. This could be done by selecting a number of time points (e.g., one every half second) and reading the value of the function from each graph. The average for all the readings for a given time slot then becomes the corresponding point on the final

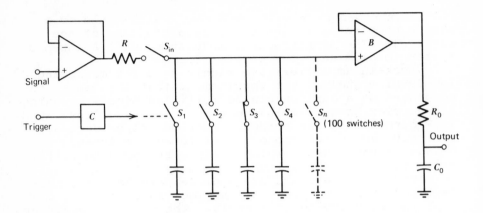

Figure 5-22. Schematic representation of a signal averager.
Switch S_3 is shown as closed. (C) is a sequential device that
closes one gate at a time in sucession. The process starts over
from the first gate each time a trigger pulse is applied.

curve. Digital computers can be used to carry out such
computations, but there are also electronic analog devices that perform the same operation. Examples are
the signal averager, and the boxcar integrator.
     The *signal averager* (Figure 5-22) consists of a
series of channels, for example 100, each containing a
capacitor and a switch. A timing circuit actuates the
switches in sequence, always beginning at a fixed time
after the trigger signal has been applied. Because of
this synchronization, on repetition, each channel will
be fed information at the same phase of the signal.
For example, channel 56 might be connected 56 msec
after each triggering. As a result, the capacitors
charge through R each time by a small amount, and after
a number of repetitions they will contain the average
corresponding to a point on the response curve. The
effect is similar to that achieved by manual averaging,
with the difference that the information is now stored
in memory by the bank of charged capacitors. To re-
trieve the information, the switch S_{in} is opened, and
a slow sequencing is undertaken. A recorder connected
at the output will register the voltages of successive
capacitors and produce the average curve. The pair
$R_{out} C_{out}$ smoothes the output, which is originally in
the form of a succession of steps from one channel to

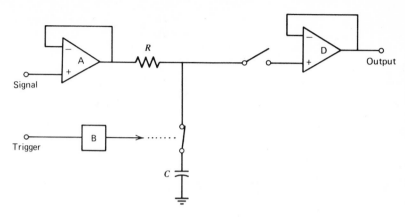

Figure 5-23. A box-car integrator.   (B) is an adjustable time-delay circuit.

the next.   Even if the result of a single run may seem
to consist mostly of noise, an average of, say, 200
runs may be quite clean looking.

A somewhat simpler averaging device is the *boxcar*
integrator, which instead of having many channels in
simultaneous operation, has only one channel (Figure
5-23).   The trigger pulse is used in this case to delay by a time interval the closing of the switch.   The
capacitor thus takes an average of a set of readings
for one single delay point.   The process is repeated
with increments in delay time until all the points are
averaged.   The operation is clearly much slower and
less efficient than the averager, but for fast signals
the increased time may not be objectionable.   Both instruments have outstanding *S/N* improvement capabilities.

GROUNDING AND SHIELDING

Most circuit schematics have numerous ground connections indicated by the conventional symbol, that are
given very little thought.   Yet inadequate grounding
is one of the major sources of improper instrument operation, especially in low-level systems.

By definition, all ground terminations must be
connected together.   This guarantees that all currents
can find a way to return to their respective sources.
It also ensures a common voltage reference, a zero
level.   These connections, however, cannot be made in

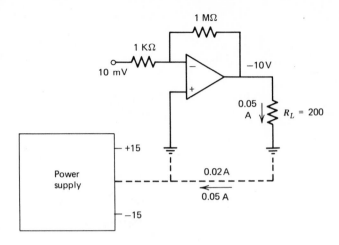

Figure 5-24. The effect of improper grounding. An input error
of 1 mV (10%) is produced. Note that the resistance of a 10-cm
length of a No. 28 copper wire is about 0.02 Ω.

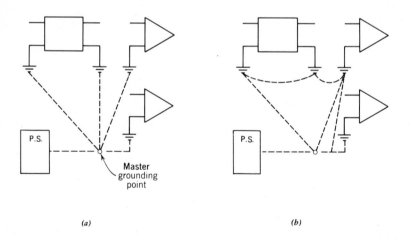

Figure 5-25. (a) A grounding system with all ground connections
brought to a single point in a star configuration. (b) An impro-
perly designed system, including undesirable ground loops.

a random way without the likelihood of impairing the precision of measurement. An example of improper grounding is shown in Figure 5-24, where the signal ground has simply been soldered onto the output return line on its way back to the power supply. If the common section of the wire has a resistance of 0.02Ω and carries a current of 50 mA, an error of 1 mV will appear, about 10% of the signal voltage.

To avoid grounding problems, two rules must be obeyed:

1. The ground wires should be as short as possible, of heavy gauge, and well soldered.
2. Each ground line should be connected by a single wire to *one* master ground point. Whether this point is actually connected to the earth potential is less important. The whole ensemble should be like a star, with no branching or looping (Figure 5-25). In low current lines, branching is less objectionable.

If the star arrangement is not practical because of the large number of ground connections, a compromise can be made using one master center point to which several heavy wires are connected, leading, for example, respectively, to (1) the power-supply common, (2) the reference inputs of op amps, (3) the ground terminals of any filters and voltage dividers, (4) any devices carrying high currents (relays, lamps, etc.), and (5) the chassis and the third line of the power cord.

Even for properly grounded systems, interference from external fields may still be strong. If the impedances involved are very large, electrostatic fields can be especially troublesome, since they consist of high voltages originating in very high-impedance sources. Fields of tens of volts per centimeter may easily be present, either from meteorological sources or simply from rubbing one's shoes on an insulating carpet. Electromagnetic fields generated by transformers, fluorescent lamps, radio stations, and the like, induce currents in conductors, which in turn generate voltage drops. In general, one can assume that stray fields are apt to become important whenever impedances larger than 1 MΩ are involved.

Interferences from these sources can be minimized by the following rules, which are particularly important in low-level measurements:

Analog Instrumentation

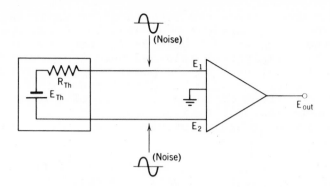

Figure 5-26. A differential system. Noise, assumed to be AC of
the line frequency, enters into both inputs with the same phase
(common mode) and is largely rejected.

    1. The AC power section of an instrument should
be physically remote from the signal section.
    2. Any line carrying high current, especially AC,
should have the two conductors twisted together so that
electromagnetic fields largely cancel out.
    3. High input-impedance amplifiers should be con-
nected, if possible, to low-impedance sources.
    4. Differential input devices should be employed
wherever practicable, to eliminate common-mode noise.
    5. Shielding and guarding should be incorporated
where appropriate.

    The principle of differential operation is shown
in Figure 5-26. The amplifier responds to the difference $E_2 - E_1$. If the noise is equally impressed on
both inputs (common mode), it tends to be subtracted
out. The ability of an amplifier to eliminate such
interferences is measured by its CMRR.
    Specially constructed integrated circuit ampli-
fiers, such as the Analog Devices AD522 or the Burr-
Brown 3626, have a well-balanced differential input
with the very high CMRR of approximately 110 dB, together with high input impedance. The feedback is internal, and the gain (1 to 1000) is controlled by a
single external resistor. These units, called *instrumentation amplifiers*, are recommended for operation in
noisy environments and for floating measurements such
as bridge readouts (Figure 5-21).

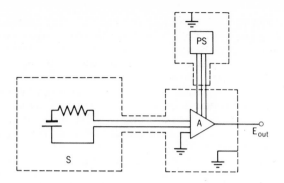

Figure 5-27. A shielding system. The source, *S*, and amplifier, *A*, have a common shield, connected at one single point into the ground system. The power supply, *PS*, has its own shield. The only ground connection between the two shielded segments is through the common line of the power-supply output.

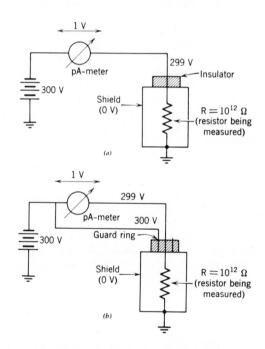

Figure 5-28. Resistance measuring devices for very high values, using (*a*) a simple shield, and (*b*) a shield with a guard.

Another type of differential amplifier is the *isolation amplifier*, which has the unique property that its input and output grounds are rigorously isolated from each other. These are used extensively in monitoring hospital patients.

Shielding consists of surrounding either the source of interference or the sensitive stages of an instrument by metal enclosures. The enclosure can be made of copper sheet or screen. The shield provides a low-impedance path to ground for various currents induced by the stray fields. It is advisable to shield the power section separately from the rest (Figure 5-27). The shields must be connected to the master ground only at one point. Shielding should extend to the connecting wires, which usually consist of one or more conductors inside a braided metal shield.

Shielding against magnetic fields requires special materials such as "mu-metal." This is inconvenient and expensive to fabricate and is used only in special cases.

In conductors carrying very small currents, leakage to the shield may produce significant errors. In this case the *guard* method can be used. The guard is a second shield connected to a voltage close to the signal level rather than to the ground. A low-impedance source must be used to power the guard. In Figure 5-28 an example is given of a device for measuring high resistances. The 300-V source generates a current through R, which is measured by the meter. The resistance in ohms is given by $R = 300/I$, where I is the current in amperes. In Figure 5-28*a* the insulator between the shield and the measuring line is subjected to practically the whole 300 V. The leakage produced by such a high voltage across the insulator can cause significant errors. In contrast, in Figure 5-28*b* the guard is within 1 V of the signal path. Any leakage between the shield and the guard returns directly to the power supply so that the error in measurement is essentially eliminated.

PROBLEMS

5-A. A description of a spring-and-weight system (Figure 5-29) must include the effect of friction, which is a force proportional to the velocity,

dx/dt, and of the acceleration, d^2x/dt^2, as well as the spring force, kx. The overall equation of motion becomes

$$m \, \frac{d^2x}{dt^2} \; = \; -kx \; - \; L \, \frac{dx}{dt}$$

Show how this equation can be simulated with op amps.

5-B. Design a circuit to solve the equation describing radioactive decay

$$\frac{dx}{dt} \; = \; - \; kx$$

where x is the amount of active isotope remaining at time t sec, and k is the decay constant. Let $k = 0.4 \times 10^{-6}$ sec^{-1} and $x_0 = 2.0 \times 10^{18}$ atoms.

5-C. A function multiplier can be designed based on the algebraic identity $(x + y)^2 = x^2 + 2xy + y^2$, provided that squaring circuits are available. Draw a diagram for such a multiplier.

5-D. Show that, in an averager such as that of Figure 5-22, after an elapsed time of RC sec a capacitor becomes charged to 63% of its steady-state value.

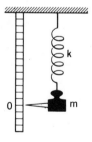

Figure 5-29. See Problem 5-A.

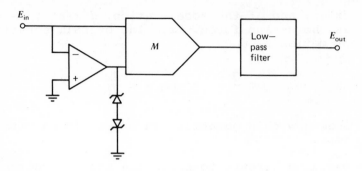

Figure 5-30.   See Problem 5-G.

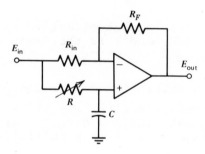

Figure 5-31.   See Problems 5-H and 5-I.

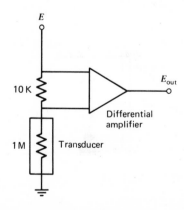

Figure 5-32.   See Problem 5-K.

5-E.   Show that a divider can be made by connecting a
        multiplier in the feedback of an op amp.

5-F.   An AC signal of 2-V amplitude and 100-Hz fre-
        quency is multiplied by a signal of 1000 Hz, also
        2-V amplitude.

        (a)   Sketch the resulting waveform.
        (b)   Sketch the waveform obtained if a 2-V DC
               bias is added at the 100-Hz input.
        (c)   Do the same for 2-V bias applied to both
               inputs.

5-G.   Figure 5-30 shows a multiplier circuit that could
        be used for perfect rectification. Explain how
        it works.

5-H.   Explain the functioning of the phase shifter of
        Figure 5-31.

5-I.   Design in detail the circuit of Figure 5-20, us-
        ing op amps, a multiplier, and the phase shifter
        of Figure 5-31.

5-J.   Draw all the ground connections for the circuit
        in Problem 5-I. Include the power supply common
        and the line ground, and show how you have avoid-
        ed ground interactions.

5-K.   In the circuit of Figure 5-32, if E = 20 V, E_{out} =
        200 mV, whereas if E = 10 V, E_{out} becomes 98 mV.
        Calculate the CMRR of the amplifier. Would this
        make a satisfactory instrumentation amplifier?

5-L.   Explain the circuit in Figure 5-33, and give the
        overall equation. What use could you find for
        it?

5-M.   In what applications would an operational trans-
        conductance amplifier be preferable to a conven-
        tional op amp? Information can be found from
        manufacturer's literature, for example, that on
        the RCA model 3060.

5-N.   Two substances, x and y, with overlapping opti-
        cal absorption spectra, can be determined simul-

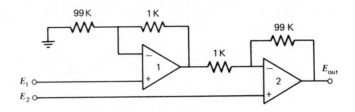

Figure 5-33.   See Problem 5-L.

taneously by means of the following equations, where A and B represent the absorbances of the mixture at two wavelengths and the a's are the absorptivities (specific constants of the system):

$$A = a_{11}x + a_{12}y$$

$$B = a_{21}x + a_{22}y$$

Show how the calculation for x and y can be carried out with an analog computer.

*   *   *

5-1.   Find the differential equations solved by the circuits of Figure 5-34.

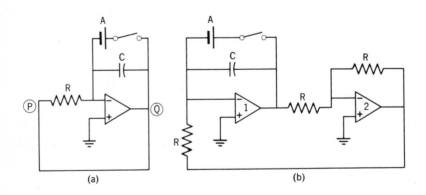

Figure 5-34.   See Problem 5-1.

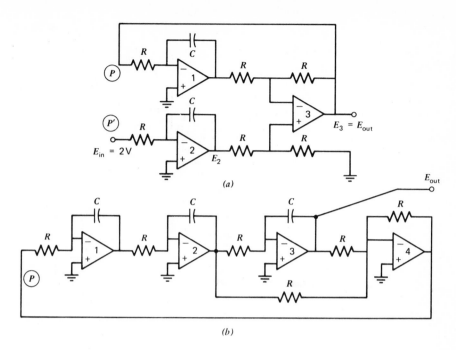

(a)

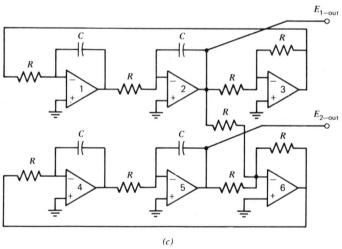

(b)

(c)

Figure 5-35. See Problem 5-2.

135

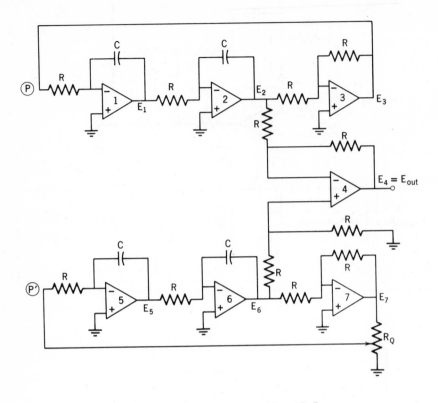

Figure 5-36. See Problem 5-5.

5-2. Write the differential equations corresponding to
      the circuits in Figure 5-35, where all R's are
      1 MΩ and all C's 1 μF.

5-3. Set up analog circuits to solve the following
      equations:

   (a) $\dfrac{d^2x}{dt^2} = -\dfrac{6}{7}\dfrac{dx}{dt} + \dfrac{1}{2}x$

   (b) $\dfrac{d^2x}{dt^2} = -10.62x + 6$

5-4. Referring to the sine-wave oscillator of Figure
      5-1b, indicate the output in both phase and am-

plitude for the following sets of initial conditions (IC):

(a) IC_1 = 0 V IC_2 = 2 V

(b) IC_1 = 2 V IC_2 = 0 V

(c) IC_1 = 2 V IC_2 = 2 V

(d) IC_1 = 0 V IC_2 = 0 V

5-5. Describe qualitatively the waveform of the output
voltage (E_4)   for the two extreme settings of
potentiometer R_Q in Figure 5-36.

VI

POWER CIRCUITS

POWER SUPPLIES

All electronic circuits, in order to operate, require a source of power, usually some DC voltages in the 5 to 30 V region. A large number of analog integrated circuits require plus and minus 15 V, whereas digital devices often need a single source of +5 V.

Power can be provided by batteries, a variety of which are available commercially. These are especially useful for portable instruments. In place of batteries, however, one usually uses *power supplies*, devices that convert the 60-Hz line power into appropriate DC voltages. Modular, plug-in power supplies are very convenient and moderately priced, but sometimes it is advantageous to assemble one's own unit.

The principal requirements of a power supply, in addition to its current and voltage ratings, are (1) absence of residual AC (ripple) in the output, i.e., good filtration; (2) an output voltage that is independent of variations in the line voltage and in the load current (good regulation); (3) a means of limiting the current to a safe value (short circuit protection).

The major parts of a power supply are shown in Figure 6-1. The fuse in the input provides short circuit protection. Simpler units might lack some of the features shown. We shall discuss the various component parts in turn.

139

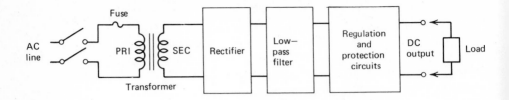

Figure 6-1. A typical power supply, block diagram.

TRANSFORMERS

A transformer (Figure 6-2) in its simplest form is a device consisting of two inductors, coupled through their magnetic fields. If the primary coil is energized by AC, the secondary will generate a voltage de-

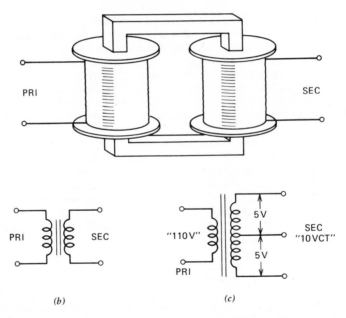

(b)   *(c)*

Figure 6-2. A transformer (a) and its conventional symbol (b). At (c) is shown a transformer with a center-tapped secondary. The more common construction has all coils located on the same leg of the iron core, but this introduces capacitance between windings, and thus a path for noise.

pending only on the primary voltage and on the ratio of
the number of turns in the two coils. The transformer
for power supply use is a heavy element, usually the
heaviest part of an instrument.

RECTIFICATION

Rectification is the first step in the conversion
of AC into DC. This can be done by inserting diodes
between the transformer and the load, as in Figure 6-3.
The current produced is limited to one direction of
flow. This gives a pulsating current that is a combination of DC and AC components. The AC ripple is very
pronounced in half-wave rectification (a, b), and less
so in full-wave (c, d).

These circuits are useful as such for a few applications, such as battery chargers, but in most cases
the ripple is objectionable and filtration is needed.

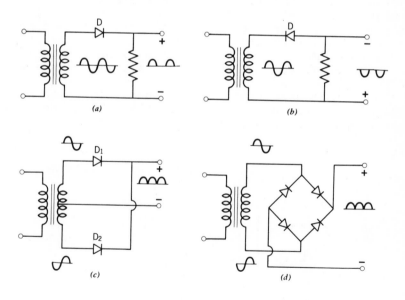

Figure 6-3. Rectifier circuits: (a, b) half-wave, and (c, d)
full-wave. The bridge (d) is available commercially as a **four-**
terminal module.

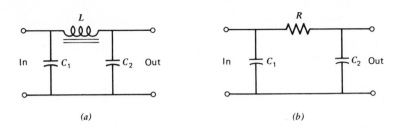

(a)   *(b)*

Figure 6-4.   Two types of power-supply filters.   (a) *LC*, (*b*) *RC*.
Both are called "π-sections" because of their diagrammatic shape.

FILTRATION

The output from the circuits of Figure 6-3 can be
fed into a low-pass filter to attenuate the AC ripple,
as shown in Figure 6-4.   The *LC* circuit in (*a*) is very
effective, but requires a bulky and relatively expensive inductor (choke).   It is particularly appropriate
in high-power systems.   In low-power systems it is usually more economical to employ circuit (*b*) in connection with an integrated regulator.   (The regulator
serves to eliminate the ripple as well as to regulate
the voltage, as will be discussed later in this chapter.)

A typical unregulated power supply using an *RC* π-
section filter is shown in Figure 6-5.   With the indicated component values, the residual ripple is something like 0.1% of the DC voltage, which is satisfactory in most cases.   The filtering improves in propor-

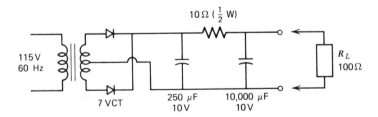

Figure 6-5.   Simple, nonregulated, power supply to give 5 V at 50
mA output.   Care must be taken to select components with high
enough voltage and power ratings.   Capacitor C_1 is kept at a relatively low value to avoid overloading the transformer.

tion to the product $RR_L C_1 C_2$. Consequently the larger
the value of R, the better the ripple rejection. On
the other hand, the smaller the R, the better the load
regulation. Observe that R and R_L form a voltage divider, hence the smaller the R, the smaller the output
voltage drop. For the example shown, under no-load
conditions, E_{out} = 5 V, whereas with a 100 Ω load, E_{out}
can be calculated to be about 4.5 V, a drop of 10%.

ZENER REGULATION

A considerable improvement for small power supplies is obtained by connecting a zener diode in parallel with the load (*zener shunt regulation*). This is
shown in Figure 6-6. It now becomes possible to use a
larger value for R, with a corresponding decrease in
the capacitors. The zener diode maintains a reason-
ably constant 5 V across the load regardless of variations in line voltage or load current. This holds
when R_L > 100 Ω, below which the diode goes out of conduction.

One factor to be taken into account in zener regulation is the power dissipated as heat in the diode.
If the load is disconnected, the diode current is limited by R to the value (10 - 5 V)/100 Ω = 50 mA. Thus
the power dissipated is $P = EI$ = 5 V × 0.050 A = 0.25
W, hence one should employ an appropriately rated diode. The power in the diode is always less when the
load is present.

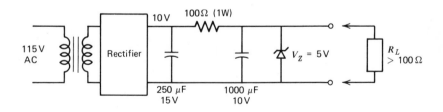

Figure 6-6. A power supply with a shunt zener diode as regulator,
designed to give 5 V and 50 mA.

Shunt regulators are wasteful of power and hence used mostly for small supplies. More effective designs will be described later in the chapter.

TRANSISTORS

In the history of electronics, vacuum tubes were dominant for many years, as the only active devices. With the development of semiconductors in the 1950s, tubes were rather suddenly replaced by transistors. The era of discrete transistors, however, was short, and nowadays integrated circuits (containing, to be sure, transistors) reign supreme. For the scientist, the use of individual transistors (and of tubes) is usually limited to circuits where the few milliamperes of output of an integrated circuit are insufficient for the purpose at hand.

BIPOLAR TRANSISTORS

The bipolar transistor is a type of three-terminal active semiconductor device. The three terminals, called *emitter*, *base*, and *collector*, represent connections to three adjacent layers or regions of semiconductor material (silicon or germanium). The semiconductor is "doped" with (contains) small concentrations of added elements that gives it either a *p* or an *n*

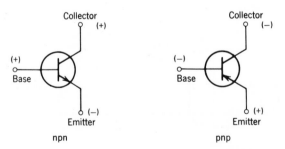

Figure 6-7. Diagrammatic representations of transistors. The arrow indicates the direction of flow of conventional positive electricity.

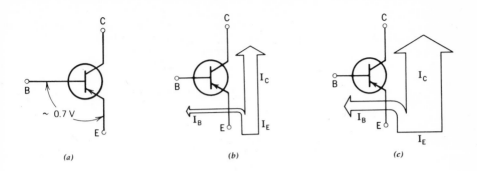

Figure 6-8.   Representation of the voltage (a) and the current flow (b, c) in a transistor. Note that the ratio I_C/I_B is the same for large or small total currents. The ratio β for various transistors lies in the range of 50 to 200.

character, that is, either positive or negative charges predominate in the transport of electricity.   There are two types of bipolar transistors, *pnp* and *npn*, the succession of letters indicating the character of the emitter, base, and collector, respectively.   The major practical difference between the two types is in the polarity of the voltages required to power them.   The schematic representations and polarities are shown in Figure 6-7.

The functioning of transistors can be understood by means of the following rules:

1.   The voltage between base and emitter is actively maintained equal to the forward voltage drop of a diode (Figure 6-8*a*):

$$E_{BE} \cong 0.7 \text{ V} \tag{6-1}$$

2.   Any current I_B fed to the base causes a collector current I_C which is greater than I_B by the factor β:

$$I_C = \beta I_B \tag{6-2}$$

where β is a constant, the current gain (Figure 6-8*b* and *c*).   This is called the *transistor effect*.

To take advantage of the transistor effect, by which a large current is controlled through the action

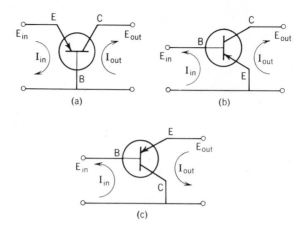

Figure 6-9. A *pnp* transistor in (*a*) common-base, (*b*) common-emitter, and (*c*) common-collector configurations. The directions of the curved arrows indicates the flow of positive current.

of a small one, the transistor must be supplied with power (and connected in a circuit with other components). In general, of the three terminals of a transistor, one acts as input, one as output, and one is utilized in common by both input and output circuits. This implies that three configurations are possible, corresponding to a common base, a common emitter, and a common collector.

Figure 6-9*a* shows a *pnp* transistor in the common-base configuration, sometimes called "grounded base." In this circuit the current amplification is always less than unity. This arrangement is often useful in very high-frequency work.

If we connect a transistor with the emitter lead common to input and output circuits (common-emitter or grounded-emitter configuration), the signal can be applied to the base (Figure 6-9*b*). By controlling the base current, it is possible to control the flow between the emitter and collector, which is β times larger.

A parameter, useful in the common-emitter configuration, is the input impedance, Z_{in}, which is conventionally defined for a transistor with short-circuited output. Typically Z_{in} is a few kilohms.

Finally, one may consider the case where the collector is connected to the common line (common- or grounded-collector configuration). In this case the signal is applied at the base while the output is taken between the emitter and common (Figure 6-9c). The current gain, I_E/I_B, is nearly equal to β. The main advantage of this configuration is that it has a large input impedance (typically 200 kΩ), and an output impedance as low as a few hundred ohms. The voltage gain is slightly less than unity, but a transistor used in this configuration can drive a device requiring large currents, and at the same time be driven by a modest current.

PRACTICAL TRANSISTOR CIRCUITS

Transistor circuits can be made to perform almost any electronic task, but nowadays integrated circuits can perform the same tasks more conveniently and economically. Nevertheless, transistors can be used to advantage when high-frequency, high currents or high voltages are involved. They are also useful for performing simple tasks, since they have less stringent power supply requirements. In this section we examine a few basic transistor circuits.

An example of current amplification, implemented with a transistor, is shown in Figure 6-10a. The operating current required by the relay is 10 mA, and the transducer (a contact thermometer) is capable of carrying safely only 0.1 mA. Amplification is provided between the base current passing through the thermometer and the emitter current needed for the relay. Their ratio (100) is approximately equal to the β of the transistor. Alternatively, a higher resistance relay could be connected in the collector circuit, as in (b). Now the base is nearly at ground potential, since the emitter is grounded. The 300 kΩ resistor connecting the base to the power supply limits the base current to 0.1 mA. The collector current is β times larger, 10 mA. The situation is the same as before, with the exception of a higher voltage across the relay (10 mA × 2 kΩ = 20 V).

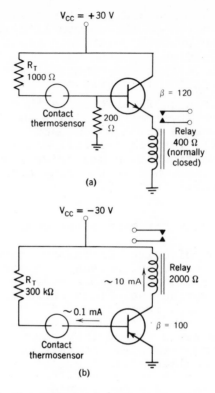

Figure 6-10.   (a) The use of a common-collector circuit to drive
a relay for a thermoregulator.   (b) A relay connected in the collector lead of a common-emitter circuit.

Among the analog uses of transistors, of special
importance for the scientist are the so-called *emitter-
followers*.   These common-collector circuits are useful
as boosters to increase the current capability of various integrated circuits.   The net effect is to reduce
Z_{Th} without affecting E_{Th}.   The basic circuit is shown
in Figure 6-11.   The output voltage is some 0.7 V smaller than the input, but the current capability is increased by the factor β.   If still larger current gains
are needed, a combination of transistors can be used.
In one such circuit, known as the *Darlington* connection, the base current of a second transistor is

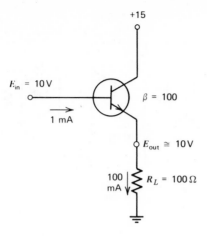

Figure 6-11. An emitter-follower circuit.

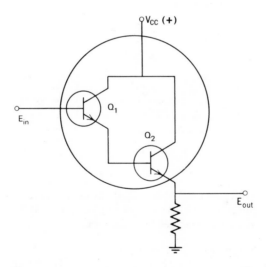

Figure 6-12. Circuit using the Darlington configuration. Such pairs are commercially available.

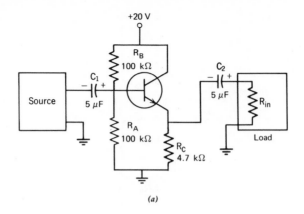

(a)

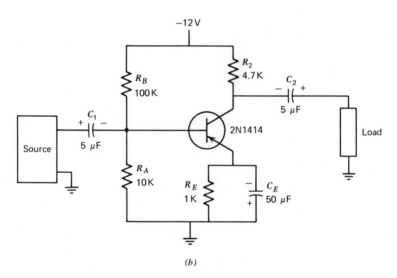

(b)

Figure 6-13. Single-transistor amplifiers. (a) Emitter-follower
or current amplifier. (b) Voltage amplifier. The capacitors are
electrolytics with polarities as indicated.

provided by the emitter current of the first.   The
combination can be treated as a single transistor (Figure 6-12). The overall β is equal to the product of
the individual βs.
       An example of the use of a follower in an AC cir-
cuit is given in Figure 6-13a. Here R_B and R_A serve to
establish a DC level, upon which is superimposed the AC

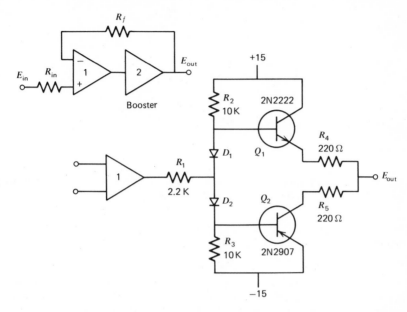

Figure 6-14. Circuit of a bipolar current booster, with typical resistor values for a 50 mA output. The diodes are type 1N4448 or 1N915. The inset shows feedback connections.

input, so that the alternating voltage will always be above ground potential. This circuit is useful for low-impedance sources feeding low impedance loads. One could use an op amp for the same purpose, except at high frequencies.

Another AC circuit, in which the output is taken from the collector, is illustrated in Figure 6-13b. In the absence of C_E, the gain would be given by E_{out} = $-E_{in}(R_C/R_E)$, reminiscent of an operational amplifier. Capacitor C_E serves to lower the impedance of the emitter circuit so that it acts as a high-pass filter.

For higher currents than an amplifier can provide, one can include a follower in the feedback loop, thereby making it effectively a part of the amplifier, and thus (ideally) error-free. This type of circuit is called a $booster$. It requires two followers, both npn and pnp, for bipolar operation (Figure 6-14). The network formed by resistors R_1, R_2, and R_3 and the diodes

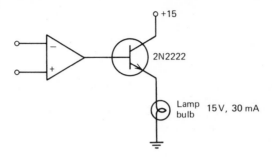

Figure 6-15. Comparator with booster, driving an incandescent lamp. The lamp is on when $E_A < E_B$.

serves to set the proper operating points for the transistors. Resistors R_4 and R_5 provide short-circuit protection as well as improved transient performance.

If only unipolar operation is needed, as for example, for driving an incandescent lamp, a booster can be constructed with only one transistor (Figure 6-15).

TRANSISTOR REGULATED POWER SUPPLIES

The shunt regulated power supply described earlier in this chapter suffers from the limitation that high-power zener diodes are expensive. A combination zener-transistor can provide a very effective, inexpensive shunt regulator, as shown in Figure 6-16. The zener current is amplified by the β of the transistor (assumed to be 50), so that a 1-A current can be controlled by an inexpensive 20-mA zener. An additional merit is that the AC current of the capacitor C_2 is also amplified by β. The circuit behaves as if the combination of C_2 and Q is a capacitor of $\beta C_2 = 50 \times 1000 = 50,000$ μF.

As mentioned previously, the shunt regulated sup ply is wasteful of power. More efficient is the series regulator shown in Figure 6-17. The transistor oper‐ ates as an emitter follower, reproducing the voltage of the zener at the output. The circuit is actually a three-terminal regulator. More complex modular three-terminal regulators are available commercially, and normally contain both short-circuit and overheating

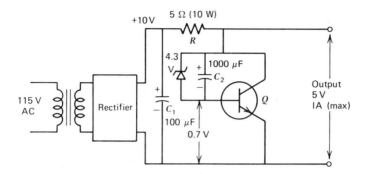

Figure 6-16.   An amplified zener shunt regulator.   The combination of Z, C_2, and Q can be thought of as a single, two-terminal unit that will carry a maximum current of 1 A.   The power dissipation in R is 5 W, and Q is also 5 W at its maximum.   Proper heat dissipation precautions must be taken.

protection.   We highly recommend them, since they are moderately priced and yet give a precision within about 0.1% in terms of regulation, drift, and ripple rejection.   A typical application is shown in Figure 6-18.

   In Figure 6-19 is given a circuit using an op amp and a transistor booster to generate a ±15 V supply suitable for powering precision op amp circuits.   The feedback guarantees that the midpoint between the resistors and hence between the positive and negative outputs is at ground potential.

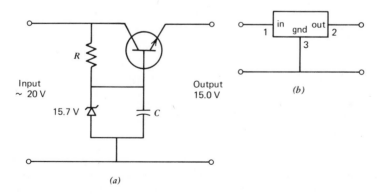

Figure 6-17.   Series regulator (a), and its black-box representation (b).   The capacitor C is effectively multiplied by the β of the transistor.

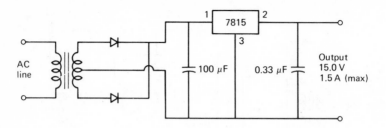

Figure 6-18.   Regulated power supply using a 7800-series modular
three-terminal regulator.   The last two digits of the 7800 number
give the output voltage.

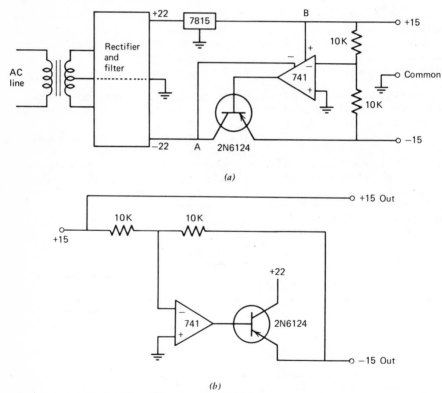

Figure 6-19.   (a) A dual polarity "tracking" power supply.   The op
amp is itself powered from −22 V at point A and +15 V at point B.
The 10 kΩ resistors must be precisely matched.   (b) A rearrange-
ment emphasizing the similarity to a unity-gain inverter.

CURRENT SUPPLIES

In the power supplies presented so far, the output voltage is held constant regardless of the current drawn by the load up to the point of saturation. In other words, by Ohm's law ($E = IR$), since E is maintained constant, it follows that the current is determined only by the resistance of the load. This is valid up to a maximum called the *current compliance* of the supply. An alternative possibility is to keep the *current* constant. In this case it is the *voltage* that is determined by the resistance of the load. The maximum available voltage is called the *voltage compliance.*

The Thevenin equivalent of a good current source consists of a high E_{Th} coupled to a high R_{Th}, such that $I_{out} = E_{Th}/R_{Th}$. There are several ways of implementing a current source, one of them being actually constructing a "Thevenin" unit. For example, a 510-V battery and 510-kΩ resistor form a nearly ideal 1-mA source, which, however, has a compliance of 510 V, something of a hazard.

Active current sources with smaller compliance can be made with op amps, an example being given in Figure 6-20. The current passing through the load must go to ground through R. On the other hand, the voltage across R is forced by the feedback to be equal to E_{in} Thus by Ohm's law the current in R, and hence in the load, is given by $I = E_{in}/R$, a constant value for given E_{in} and R. Furthermore, the supply is programmable,

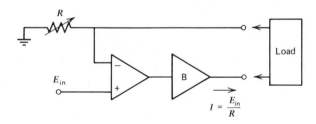

Figure 6-20. An adjustable constant-current supply. Amplifier B is a booster within the feedback loop. Note that the load must be isolated from the ground.

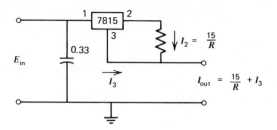

Figure 6-21. Simple current supply using a voltage regulator similar to that in Figure 6-18. I_3 is a constant current flowing from terminal 3 (typically 5 mA).

meaning that one can set any desired current by providing an appropriate R. A very simple constant current supply can be made with a modular voltage regulator (Figure 6-21).

THYRISTORS

Thyristors are semiconductor components with the distinguishing characteristics that they can be turned on (rendered conductive) by a suitable signal, but removal of that signal fails to turn them off. They may thus be considered the semiconductor analog of a relay that latches in the on position.

The basic type of thyristor is a three-terminal device called a *silicon-controlled rectifier* (*SCR*). Its characteristics follow the curves of Figure 6-22.

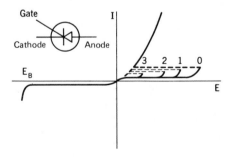

Figure 6-22. The current-voltage characteristic of an SCR. Curves 0 to 3 represent increasingly positive gate voltages.

The general behavior is that of a conventional silicon
diode, except for the region corresponding to moderate
forward currents. Let us assume first that no connection is made to the gate electrode. If one now increases the anode-cathode potential in the forward direction (curve 0), the current will at first assume a
very small value (a few miliamperes in an SCR capable
of handling tens of amperes). It will follow the horizontal curve for a considerable voltage range, and
then turn upward. At this point a forward "breakover"
occurs (dotted line) and the device reverts to normal
diode operation, with the current limited only by the
external circuit resistance. At the same time, the
potential drop across the unit falls to perhaps 1 V.
This abrupt decrease in impedance is called "firing" or
"turning-on" of the thyristor. Once fired, the device
cannot be turned off again except by interruption of
the anode current. The breakover voltage can be de-
creased by supplying a positive potential to the gate
relative to the cathode. Curves 1 to 3 correspond to
progressively larger gate voltages. The device is
operated in the region below the spontaneous breakover,
so that a gate signal is required for firing. Thus the
unit acts as an open switch until it receives a positive pulse at the gate, when it is converted to a closed switch.

There are a number of types of thyristors in addition to the SCR. One is the *triac*, which is a bidi-
rectional modification of the SCR; its characteristic
curve is the same as for the SCR in the first quadrant,
but repeats that curve symmetrically in the third quadrant (Figure 6-23).

The flexibility of thyristors arises from the
many circuits available for turning them on. The simplest of these involves an AC-triggering signal. Figure 6-24 depicts a typical case—an emergency device to
supply battery power to a lamp automatically if the AC
line fails. With AC power on, the capacitor charges
through diode D_1 and resistor R_1 to develop a negative
voltage at the gate of the SCR. By this means the SCR
is prevented from triggering, and the emergency light
stays off. Should the AC power fail, the capacitor
discharges and the gate becomes positive by the action
of the battery through resistor R_3. Diodes D_1 and D_2
are now reverse-biased and play no role. The SCR then

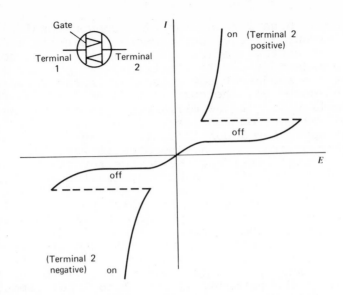

Figure 6-23. The triac thyristor.

fires and energizes the lamp, the DC current passing
through the secondary of the transformer. Reset is automatic when AC is restored. The function of D_2 and R_2
is to maintain the battery fully charged.

Probably the most useful turn-on circuit involves
a *unijunction transistor (UJT)*. The important charac-

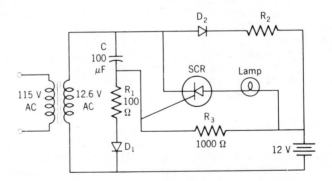

Figure 6-24. Storage-battery operated emergency light, using an
SCR. A variety of loads could be substituted for the lamp. (General Electric Company).

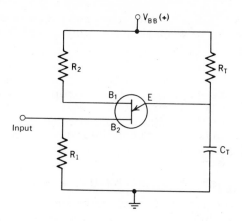

Figure 6-25. Pulse generator using a unijunction transistor.

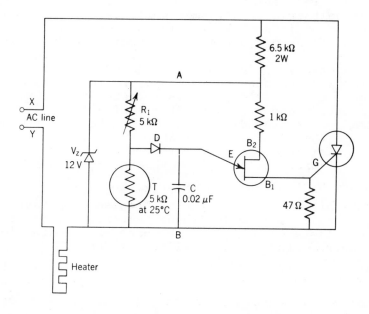

Figure 6-26. A temperature control circuit using UJT and SCR in combination (General Electric Company).

teristic of the UJT is that the resistance between two
of its terminals, the emitter E and first base B_1, suddenly decreases from its normal value of several kilohms to nearly zero when the potential of the emitter
is raised above a critical value, thus permitting rather large currents to pass. This property allows the
design of a simple circuit for generating a train of
pulses (Figure 6-25). When connection to the power
supply is made, the capacitor begins to charge through
R_T. When its potential equals the critical value, the
resistance between E and B_1 drops abruptly and the capacitor discharges through the UJT and R_1. When it
reaches some low level (which depends in a reproducible
manner on the magnitude of R_1), the base-emitter resistance returns to its original high level and the capacitor begins to recharge, starting with the next cycle.
        Figure 6-26 shows an application of a UJT to con-
trol an SCR in a temperature regulator. This circuit
contains neither a transformer nor a relay, hence can
be constructed in a conveniently compact form. The
condition of the circuit must be studied separately for
successive half-cycles of the AC power. When the lower
line connection, Y, is positive, the SCR cannot conduct,
no matter what happens to its gate, while the zener
diode can conduct freely in its forward direction.
Hence points A and B will be at essentially the same
potential, so that the UJT cannot operate. On the next
half-cycle the upper line, X, is positive, with the
zener maintaining A at 12 V more positive than B. The
capacitor C proceeds to charge through R_1 at a rate
determined by the resistances of R_1 and R_T, where T is
a thermistor* immersed in the bath whose temperature is
to be controlled. The lower the temperature, the larger the resistance of T, and the greater the fraction
of the current through R_1 available to charge C. This
increases the frequency of production of pulses by the
UJT. The direct connection between B_1 and G ensures
that the SCR is turned on at the time within each cycle

*The thermistor is a resistor made of metal oxides sintered to-
gether. It shows a characteristically high negative temperature
coefficient.

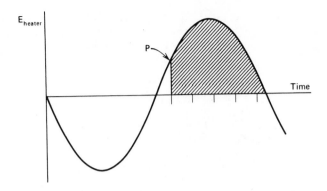

Figure 6-27.   The operation of the circuit of Figure 6-26.   The
heater is on during the shaded portions of the cycle.   The ticks
correspond to successive firings of the unijunction transistor.

when the first pulse from the UJT appears (Figure 6-27).
The SCR is turned off automatically when the voltage
crosses zero.   The energy supplied to the heater corresponds to the shaded portions of the diagram, and is
seen to be greater when the pulse rate is faster, corresponding to a temperature lower than desired.   The
temperature level can be set by adjustment of the variable resistor R_1.

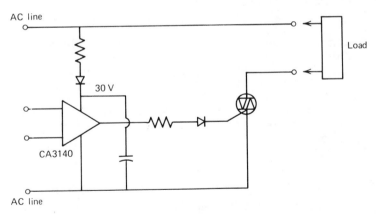

Figure 6-28.   A triac triggered by an op amp.   Note that the amplifier is powered by half-wave rectified AC.   (Courtesy RCA Corporation, File No. 957)

An interesting application of thyristors, in which an op amp is used to trigger a triac is shown in Figure 6-28. The amplifier can be connected as a comparator to trigger at a given input level.

PROBLEMS

6-A.   Consider a constant-current source consisting of a 510-V battery and a 2-MΩ resistor.

   (a)   What is the current output into a 1-kΩ load? A 10-kΩ load?
   (b)   What is the regulation, defined as $100(\Delta I/\Delta R)$? What are its units?
   (c)   What is the voltage compliance?

6-B.   Calculate the average and RMS values for the output voltages in the circuits of Figure 6-3a and c.

6-C.   Consider Figure 6-14 with the following parameters: R_{in} = 10 kΩ; R_f = 10 kΩ; E_{in} = -2 V (DC); the β's of both transistors equal 100; the voltage drop in each diode is 0.6 V: the voltage drop between base and emitter for each transistor is 0.6 V; the load is a 1-kΩ resistor from output to ground. Calculate the voltages at all junction points in the circuit.

6-D.   Consult manufacturer's literature concerning the 7800 series of voltage regulators. Design a variable output power supply giving 0 to 10 V, capable of at least 50 mA, using one of these.

6-E.   Make a report on the uses of the RCA model CA3059 integrated circuit, used in connection with triacs.

*   *   *

6-1.   In the voltage-regulating circuit of Figure 6-29a, with E_Z = 12 V, compute the resistance and necessary power ratings of R and the maximum power dissipation in the zener, if the load may vary

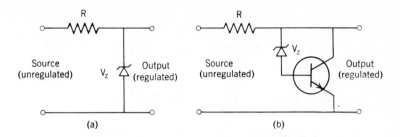

Figure 6-29. See Problems 6-1 and 6-2.

from 100 to 2000 Ω. (It may be assumed that E_Z is constant over the required range of currents. Assume E_{in} = +15 V.)

6-2. In the circuit of Figure 6-29b, with R = 100 Ω, E_Z = 10 V, and E_{in} = 20 V, what should be the power ratings of the three components?

6-3. In Figure 6-3a the diode and resistor pair can be thought of as a voltage divider. Calculate the ratio E_{out}/E_{in} for both the positive and the negative half-cycles of the AC. Use the following data: R = 1 kΩ, R_{fwd} = 1 Ω, R_{rev} = 1 MΩ. (R_{fwd} and R_{rev} denote the forward and reverse resistances of the diode. Neglect the forward voltage drop.)

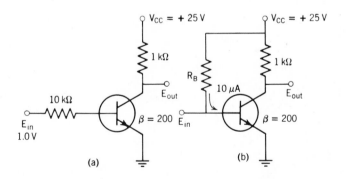

Figure 6-30. See Problems 6-4 and 6-5.

6-4. Consider the circuit of Figure 6-30a, and com-
      pute the following: (a) I_B, (b) I_C, (c) E_{out},
      and (d) A_V. Neglect the base-emitter voltage
      drop.

6-5. For the circuit in Figure 6-30b, find the follow-
      ing quiescent quantities: (a) E_{out}, (b) the
      "transimpedance," dE_{out}/dI_{in}, and (c) the value of
      R_B.

VII

OTHER ACTIVE DEVICES

The *field-effect transistor* (*FET*) is a *unipolar* transistor, in contradistinction to the types previously discussed, which are *bipolar*. The functioning of a FET can be understood by considering it hypothetically as a bar of doped silicon with contacts at both ends, and with a collar of oppositely doped material surrounding it (Figure 7-1). (Actual FETs are not shaped like this.) A lengthwise cross-section of the FET is shown in Figure 7-1*b*, in which the channel is made of *n*-type silicon, and the collar, called the *gate*, of *p*-type silicon. Of the two terminals, the *source* is made more negative than the *drain*, and the gate still more negative. If the gate-to-source voltage, V_{GS}, is zero, the channel forms a low-resistance path from drain to source. If now the gate is made progressively more negative, the electrostatic field, which it induces inside the bar, tends to push away the free electrons in the *n*-type silicon, leaving a smaller and smaller channel to conduct the current. Eventually the field will so effectively strangle the channel that no current can pass. Note that the gate-channel junction is reverse-biased at all times, which means that the gate will not draw any significant current. It follows that the voltage of the gate is a more convenient operating variable than the gate current—a point of difference between FETs and bipolar transistors.

The transistor just described is known as a junction FET (JFET). Just as with bi-polar transistors,

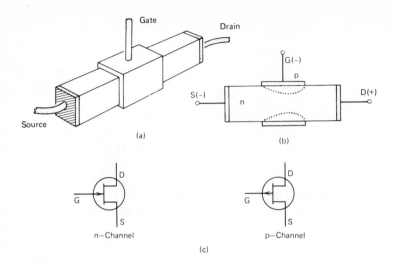

Figure 7-1. The principle of the FET (*a*, *b*), and the schematic representation (*c*). It is important to observe the correct polarity of the gate with respect to the channel, since the junction constitutes a low-power diode that could fail if forward biased.

two forms of JFET exist, with n and p regions interchanged. Figure 7-1*b* describes an n-channel JFET; a p-channel unit merely requires reversal of signs. The conventional symbols for both types are shown in Figure 7-1*c*.

Many FETs are symmetrically constructed, so that the drain and source connections can be interchanged without affecting the operation. Others are asymmetric, optimizing one of the two possible arrangements.

Another class of FETs are those in which the gate is separated physically from the channel by a thin layer of insulating material, usually silica (SiO_2).

These are called *MOSFETs* (metal-oxide-semiconductor FETs). Since the MOSFET contains no pn junction, the gate potential is no longer restricted to a single sign as in the JFETs. The gate can never draw direct current; the input resistance is even higher than for JFETs. The lack of sign restriction permits greater flexibility in circuit design.

Figure 7-2 shows typical characteristics of several types of FET. In Figure 7-2*a* is a curve for an n-channel JFET; note that the gate can only be nega-

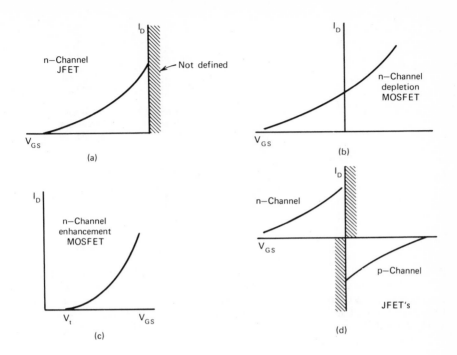

Figure 7-2. The transfer characteristics of four types of FET. The curves in (a), (b), and (c) are for n-channel units; for p-channel FET's the polarities of both I_D and V_{GS} are reversed.

In (d) a comparison is made between an n-channel and p-channel JFET. In (a) the current is essentially zero for large negative values of V_{GS}, and increases as V_{GS} becomes more positive. The region on the positive side of the graph is not defined, since JFET's cannot be used with a forward-biased gate. In (b) the portion on the left is similar to (a), but the insulated gate permits operation with a positive bias. In (c) no current will flow for V_{GS} negative, and even for small positive voltages up to a threshold V_t. The polarity relationship between the three terminals in this case is the same as in the bipolar transistor.

tively (i.e., reverse) biased. The curve in Figure 7-2b is that of an n-channel MOSFET, showing that the gate can now be made positive, which enhances the conductivity of the channel. The transistor is said to be operating in the *enhancement* mode when the gate is

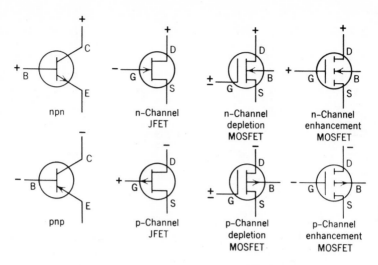

Figure 7-3. The symbolic representations of different types of FET. The bipolar transistors are shown for comparison. The fourth terminal in MOSFETs, marked "B", connects to the *bulk* of the semiconductor material. In many circuits the bulk is connected to the ground, in some MOSFETs it is internally tied to the source.

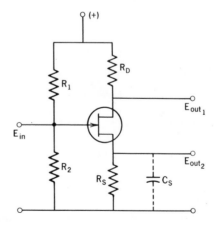

Figure 7-4. A basic FET amplifier. The transistor may be either *n* or *p* channel, but the polarity of the power supply must be selected accordingly. The capacitor C_s is useful in AC applications when the output is taken at E_{out-1}.

biased with sign opposite to that of the majority carriers (p or n) in the channel; when the sign is the
same, operation is in the *depletion* mode. (In JFETs
this is the only mode available.) Some MOSFETs operate
in the enhancement mode only; in these no current will
pass between source and drain at zero gate voltage
(Figure 7-2c).

Figure 7-3 compares the symbols and polarities for
the several kinds of FET, both n and p channel, along
with bipolar types for reference.

FETs are very useful in amplifier construction,
since they exhibit extremely high input impedance, a
desirable characteristic. One major limitation is that
having high input impedance makes them susceptible to
damage from electrostatic discharges.

A typical FET amplifier circuit is shown in Figure
7-4. Note that E_{out-1} and E_{out-2} are 180° out of phase
(i.e., when one increases, the other decreases).

Of importance in instrumentation is the ability of
a FET to act as a voltage-controlled resistor (Figure
7-5). Thus a FET can have a resistance of 50 Ω when

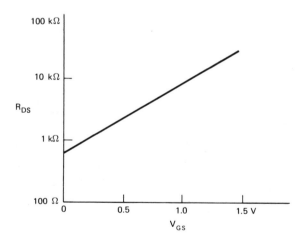

Figure 7-5. The dependence of the resistance of a FET on the
gate-to-source voltage V_{GS}. Observe that the dependence is logarithmic.

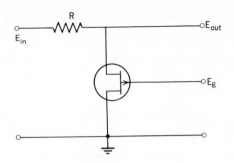

Figure 7-6. A voltage-controlled divider using a FET.

"on" and perhaps 10^9 Ω when "off." An application of
a voltage-controlled resistor is shown in Figure 7-6,
where the channel of a FET together with a resistor R,
forms a voltage divider, so that for a given E_{in} the

output can be changed over a wide range by varying the
control voltage E_G. This circuit could be used as an

automatic volume control in a radio receiver or as a
variable attenuator in a measuring instrument.
      If this circuit is doubled in the sense shown in
Figure 7-7, we obtain a Wheatstone bridge, which can be

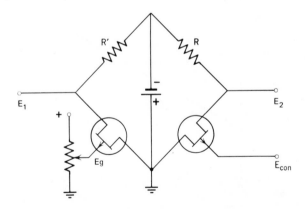

Figure 7-7. A self-balancing Wheatstone bridge using FETs. This
circuit is very similar to the first stage of a FET-input op amp,
where the inputs are connected to E_{con} and E_b.

made self-balancing. The control voltage E_{con} is to be
supplied from an electronic system that senses the difference in potential between E_1 and E_2, and reacts to
minimize this difference.

Also of great utility are the applications where
FETs are made to operate as on-off switches, as shown
in Figure 7-8. The gate in Figure 7-8a, using a single
transistor, requires a 15-V command signal, which is
inconvenient if driven from an op amp. On the other
hand, the gates in Figure 7-8 b and c require only 5 V.
In Figure 7-8c the desired analog path is selected by
an appropriate combination of voltages at the control
inputs, as was the case in Figure 4-19. Considerably
more about this type of control operation will be given
in later chapters. It suffices to note here that, in
characteristic IC fashion, one need not understand the
internal workings, if one knows the response for a giv-
en control input.

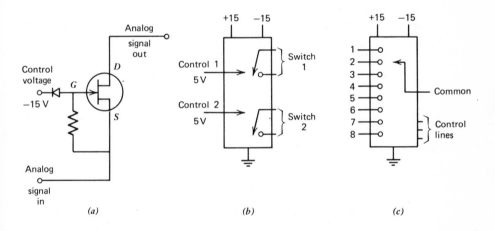

Figure 7-8. Examples of FET transmission (analog) gates. (a)
Using a discrete transistor; the diode prevents forward biasing
of the FET gate and allows the handling of analog signals of either sign. (b) A pair of on-off switches in a common IC. (c) An
8-position FET switch, known as a *multiplexer*. (A possible short-
coming of integrated multiple switches as against discrete units
is crosstalk between channels; one is tempted to say that IC
FETs are not *discreet* enough.)

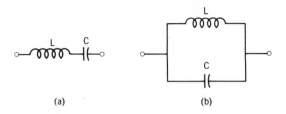

(a)   (b)

Figure 7-9. Series (a) and parallel (b) LC circuits. The impedance of (a) reaches a minimum at the resonant frequency, f_{res} =
$1/(2\pi\sqrt{LC})$, while (b) exhibits a maximum at the same frequency.

OSCILLATORS

Oscillators are electronic devices that convert
power from a DC supply to alternating current. The
output may be a pure sine wave, or it may contain harmonics in addition to the fundamental. Sources of
special waveforms, such as square, sawtooth, or triangular waves are more often called *signal generators*.
An oscillator must have some frequency-determining
network. This function may be taken by a resonant circuit consisting of a combination of an inductor and a
capacitor (a *tank* circuit). The resonant frequency
(in hertz) for an LC circuit is given by

$$f_{res} = \frac{1}{2\pi\sqrt{LC}} \qquad (7-1)$$

where L is the inductance in henries and C the capacitance in farads (Figure 7-9). The inductor and capacitor of the tank may be connected in series, in which
case the impedance is a minimum at the resonant frequency, or (more commonly) in parallel, with a resulting maximum impedance at resonance. The minimum or
maximum can be very sharp if the ohmic resistance of
the inductor is low, and if the losses in the capacitor
are small. For low audio frequencies, inductors are
too bulky for convenience, so that tank circuits are
seldom employed in this region.
Figure 7-10 shows a few LC oscillator circuits.
In each a connection is made from the tank to the base,
in such a way that its phase at the tank frequency is
180° different from that at the collector. This phase

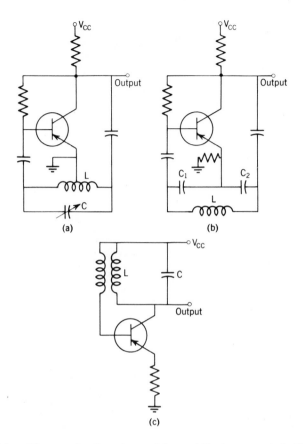

Figure 7-10. Three single-stage *LC* oscillators: (*a*) the Hartley circuit, with a tapped inductor, (*b*) the Colpitts circuit, with a divided capacitor, (*c*) a transformer-coupled oscillator. In each case the frequency is given by $f = 1/(2\pi\sqrt{LC})$.

differential matches the inherent 180° phase shift of the transistor, so that the net effect is one of positive feedback, which increases the amplitude at frequency f. After a short time the only frequency present will be f, and the circuit is said to oscillate at this frequency.

Another variety of oscillator is shown in Figure 7-11. It employs a piezoelectric quartz crystal as the frequency controlling element. This oscillator is similar to the AC amplifier in Figure 6-13*b* with the com-

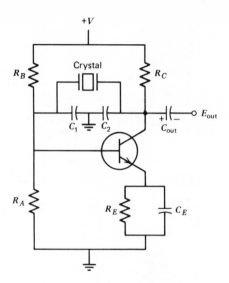

Figure 7-11. An example of a crystal oscillator. This is called the Pierce circuit.

bination of crystal and capacitor connected as feedback between output and input. Two capacitors are needed to

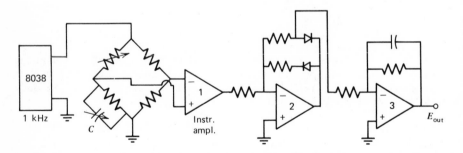

Figure 7-12. An AC bridge and associated circuitry. The capacitor C is needed for nulling the phase of the bridge current. Such a bridge is useful in connection with strain gauges, thermisters, electrolytic conductance, and other resistive measurements. The bridge is first zeroed both resistively and capacitively; the output then measures the subsequent unbalance of the bridge with variation in the resistance of the transducer.

generate the appropriate phase relation. Crystal oscillators are among the most precise devices available,
with stabilities within the sub-parts-per-million range.
  Among various IC oscillators, a good example is
the type 8038 (Intersil, Exar), which provides simultaneously sine, square, and triangular waves. The frequency can be tuned by an RC circuit or by a voltage,
from about 1 cycle in 20 min (<1 mHz) to 10^6 Hz, with
very good amplitude stability. Such oscillators, while
having a rather high (1%) harmonic distortion, are
ideal for powering AC bridges, an example of which can
be seen in Figure 7-12. The output from the AC bridge
is fed differentially into the instrumentation amplifier. A perfect rectifier converts it into pulsating
DC, which is then filtered. Note the rather small
parts count, typical of modern IC techniques. In reality, in the four ICs, there are over 100 transistors.
  A widely used IC is the 555 timer. For example,
a 555 can be used to expand a short trigger pulse (monostable operation) as seen in Figure 7-13. The load

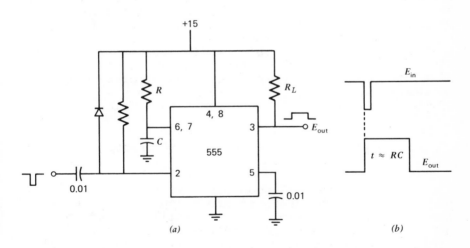

(a)   (b)

Figure 7-13. The 555 timer in the time-delay (monostable) mode of
operation. (a) The circuit, showing pin numbers. (b) The time
relation between input and output pulses; the input pulse can be
of any width without changing the output pulse, because the circuit responds not to the pulse itself but to its downward transition.

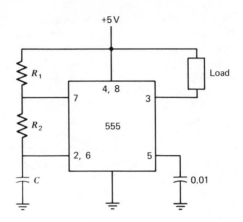

Figure 7-14. A square-wave generator using a 555 IC. R_1 and R_2 establish the symmetry of the square wave (its *duty cycle*).

could be a relay that remains on for a fixed time interval after being turned on by a pushbutton. In general, almost any short-period timing problem can be solved with such a circuit, from burglar alarms and Christmas tree lights to microsecond pulse work.

Another useful application of the 555 is as a square-wave generator, as shown in Figure 7-14. A good sine wave can be obtained from the square wave by means of a sharply tuned band-pass or even a low-pass filter to remove harmonics. There are many other applications of the versatile 555, too numerous to mention here.

OPERATIONAL AMPLIFIER OSCILLATORS

There are several possible ways in which sine-wave oscillators can be designed around op amps. One of these, using three amplifiers, was illustrated in Figure 5-1*b*. The essential requirement is a network providing positive feedback selectively at one particular frequency. Figures 7-15 and 7-16 show two other circuits, provided this time with amplitude stabilization.

In the circuit of Figure 7-15, the requisite phase shift is supplied by three stages of RC filtering, each one providing 60° of phase shift at f_0. Figure 7-16 shows the circuit of a *Wien-bridge* oscillator that con-

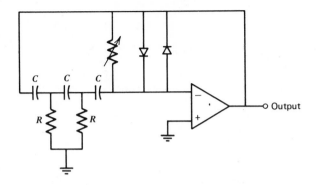

Figure 7-15. A phase-shift oscillator. The frequency is given by $f_0 = 1/(2\sqrt{3}\pi RC)$.

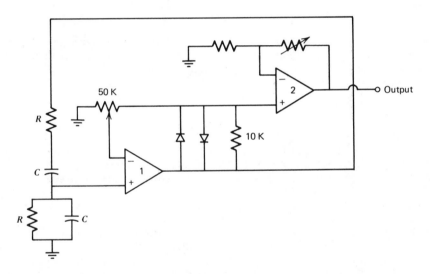

Figure 7-16. A Wien-bridge oscillator. The frequency is $f_0 = 1/(2\pi RC)$.

tains two RC combinations, one in series, the other in parallel, to determine the frequency. The output impedance of this circuit is rather high, so a voltage follower is essential. The 50-kΩ variable resistor should be adjusted to give minimum distortion. In both of these circuits, the feedback diodes serve to limit the amplitude, decreasing harmonic distortion.

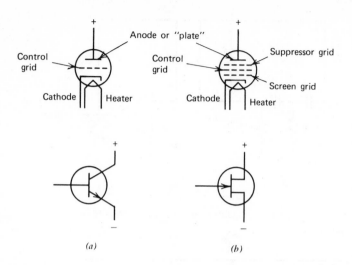

(a)   (b)

Figure 7-17. The triode (a) and pentode (b), compared with transistors. The suppressor and screen grids must be supplied with
proper voltages for optimum operation. The signal is fed into
the control grid.

VACUUM TUBES

There are still a few uses for vacuum tubes, aside
from replacements in old TV sets. High-power, high-
frequency circuits are still designed with tubes.
These units operate by an electrostatic field controlling a beam of electrons travelling through a vacuum.
The most important types of vacuum tubes are the triode
and the pentode (Figure 7-17), which can be likened to
npn transistors and n-channel FETs, respectively. The
biasing arrangements are also similar. Tubes, as compared with transistors, are larger, more expensive,
and dissipate a great deal more heat. For high-power
applications, this might not be overriding, since tubes
usually have superior frequency response and considerable immunity to electrical power transients.

PROBLEMS

7-A. Show how FETs can be used to control the "integrate," "hold," and "reset" modes of an integrator.

7-B. Design a time-delay burglar alarm using a 555
      timer. After entry, the delay allows 20 sec for
      the owner to disable the alarm by means of a se-
      cret button.

7-C. Show how one can use several 555 timers to gen-
      erate four consecutive periods of 10 sec each,
      to turn on various colored lights on a Christmas
      tree.

7-D. Suppose that in the perfect rectifier of Figure
      4-23b, the diodes are replaced by FET switches
      driven by alternate half-cycles of a square wave
      at the same frequency as the input signal.

      (a)   Can this be used as a rectifier?
      (b)   Show how it can be used to measure phase
            angles.

7-E. Look up in reference books and manufacturer's
      literature to find the actual geometry of FETs,
      and compare with Figure 7-1a.

7-F. Devise a circuit whereby a phase-shift oscillator
      can be made voltage-tunable by the addition of a
      pair of matched FETs.

                    *   *   *

7-1. Consider the circuit in Figure 7-18. Sketch the
      output for the following inputs, E_{in}:

      (a)   4 V, DC.
      (b)   4 V, RMS sine wave at 50 Hz.

7-2. Calculate the natural frequency of oscillation of
      the tank of Figure 7-10b.

7-3. Design a signal generator to give saw-tooth waves.
      [Hint: use an integrator, a comparator, and a
      FET-switch.]

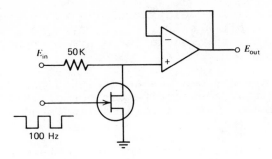

Figure 7-18.   Circuit for Problem 7-1.

VIII

LOGIC SYSTEMS

In the previous chapters the major thrust has been
the treatment of signals in the analog domain. Let us
turn now to the digital realm, where in place of amplifiers and function generators, we encounter such devices as logic gates, digitizers, and counters.

The interrelation between the two fields can be
best understood with an example, such as shown in Figure 8-1, an autoranging voltmeter. The analog input,
after buffering, is fed into an amplifier whose gain
is controlled by two logic signal lines. These lines,
in turn, are controlled by the status of the two comparators. Whenever the voltage is outside the range of
1 to 10 V, the control logic will shift the position of
the switch in the appropriate direction. The purpose
of this scheme is to permit the measurement of any DC
voltage between 1 and 9999 mV with four-digit resolution. The necessary information about the decimal
point is forwarded to the output. The analog-to-
digital converter (A/D in the figure) serves to generate the actual digits for display.

Before proceeding to further discussion of such
instruments and their components, let us devote some
time to fundamentals.

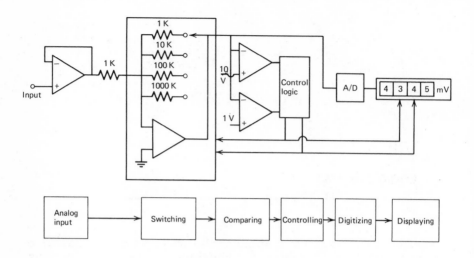

Figure 8-1. An auto-ranging voltmeter. The labels beneath the block diagram describe the functions performed. The control switch is similar to the one in Figure 4-19.

LOGIC STATES

Since every logic signal must exist in only one of two states, each logic device must give an output at either one of two levels (voltages), but nowhere in between (Figure 8-2). Similarly, each device that receives one of these signals must be able to respond appropriately to each of the two levels, but need not produce a defined response to any intermediate voltages.

The two levels are designated by many different names: *yes* or *no*, *high* or *low*, *plus* or *minus*, *true* or *false*. Perhaps the most prevalent names are "*logical 1*" and "*logical 0*," in reference to the binary system of numeration. Any logic system must have established voltage levels corresponding to these two states. The two voltages could, in principle, be of any desired values, and in fact various manufacturers have chosen differently. In the present treatment 0 V is selected for "logical 0" and +5 V for "logical 1." There is some latitude in these voltages, as indicated in Figure 8-2.

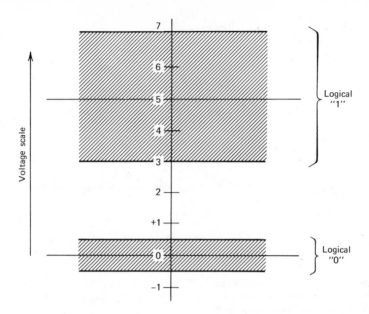

Figure 8-2. Conventional logic levels. Nominal values are 0 for logic "0" and +5 V for logic "1". The levels are not permitted to lie within the limits of +0.5 and +3 V.

ANALOG AND DIGITAL SIGNALS

A point in a circuit that can take logic levels 0 and 1 can be considered to carry the values of a special type of variable that can have only these two values. This is called a *digital variable*, and an electrical signal carrying it is a *digital signal*.

Of particular interest are groups of digital signals. Such groups can be represented by a string of ones and zeros called *bits*. For example, 0001000001 indicates the states of 10 lines, with the fourth and tenth at logical "1", and all the others at logical "0," as shown in Figure 8-3*a*. Such a group of logic signals, interpreted as a unit, is called a *word*. A digital word can also be generated as a time succession of ones and zeros, as shown in Figure 8-3*b*.

The process by which we interpret the group of logic states as a number is called *digitization*, and greatly simplifies their manipulation. Each value de-

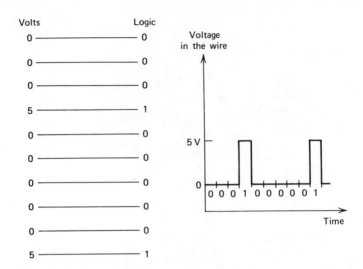

Figure 8-3. The transmission of logic words, (a) parallel, using 10 wires, and (b) in serial form. In both, the information is the same. Case (b) requires only a single wire instead of ten, but takes more time to transmit the signal.

fines a unique binary number and thus a unique combination of ones and zeros.

There are in use various coding systems that can represent letters and other symbols in addition to numbers. The most common of these is the so-called ASCII teletypewriter code. In this representation the word 0001000001 corresponds to the letter A. It is this richness of interpretations as true/false states or as numbers or letters or as voltages, that explains the extreme usefulness of digital signals. In the following sections we will elaborate on the manipulations of digital signals, regardless of the interpretation of the specific cases.

OPERATIONS ON SIGNALS

The fundamental unit in handling digital signals is the *logic gate*. A logic gate is a device in which a specific output state is generated by any given combination of input signals. In a typical logic system, there may be anywhere from a few to hundreds or thou-

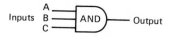

Figure 8-4. The AND gate (a) and its truth table (b) that details
the possible states of logic.

sands of individual logic gates. Each of them can be
physically small, as they are never required to handle
much power. It is therefore a great advantage to utilize modular logic components fabricated as integrated
circuits, each containing many logic elements.

   An example of such an element is the AND gate
shown in Figure 8-4. Note that the output is a logic
variable taking the value "1" only if A *and* B *and* C
are all at logic "1" (hence the name AND gate). It is
interesting to note that the output represents a reduction in the number of variables. One can say that
the AND gate answers the query: "Are all inputs at
logic '1'?," giving the response of "1" for "yes," "0"
for "no." Alternatively, one can think of it as answering the question, "Is the input the number 7?"
since, as we will see, binary 111 is equivalent to
decimal 7. The table accompanying the gate in Figure
8-3 is called a *truth table* (a name borrowed from symbolic logic). This table describes exhaustively *all*
possible input states, this in strong contrast to analog devices, where signals can have an infinity of
values.

   Now observe the related gate in Figure 8-5, called
the OR gate. In this case, the question to be answered
is, "Are any of the inputs A *or* B *or* C at logic '1'?"
Note that the answer is "1" except if A, B, and C are
all zero. Another way of looking at the OR gate is to
consider it as answering the question "Is the number
represented by the input different from zero?" As an

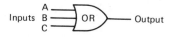

Figure 8-5. The OR-gate (a) and its truth table (b). There is
no theoretical limit to the number of inputs for AND and OR gates,
but practically, they are seldom seen with more than eight.

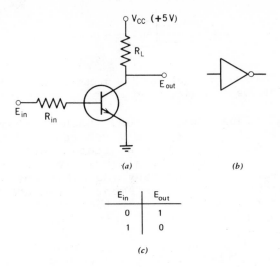

(a)   *(b)*

E_{in}	E_{out}
0	1
1	0

(c)

Figure 8-6. The logic inverter: (*a*) a circuit, (*b*) the symbol
(not to be confused with an op amp), and (*c*) the truth table.

example, if the signal inputs represent states of a
set of switches, one can think of the OR gate as giving
a "0" output whenever all switches are open, otherwise
"1."
      There are two other basic gates, which are the re-
verse of the AND and OR gates, called NAND and NOR, respectively. Each of these can be made up by combining
its prototype with an *inverter* (Figure 8-6), a circuit
element that converts a "1" into a "0" and a "0" into
a "1." The inverter is readily constructed from a single transistor, as shown, but is also available as an
IC, as are all the gates including the reversed ones.
A NOR gate, with its symbol and truth table, is shown
in Figure 8-7.
      The truth tables and symbols for each of the gates
mentioned (and an additional gate, the "exclusive-OR")
are summarized in Table 8-1. Note that inversion is

Inputs A
        B ——| NOR)o—— Output
        C

Figure 8-7. A NOR-gate (*a*) and its truth table (*b*).

TABLE 8-1

Truth Tables for Several Gates

Inputs			AND	NAND	OR	NOR	EXCLUSIVE-OR
A	B	C					
0	0	0	0	1	0	1	0
0	0	1	0	1	1	0	1
0	1	0	0	1	1	0	1
0	1	1	0	1	1	0	0
1	0	0	0	1	1	0	1
1	0	1	0	1	1	0	0
1	1	0	0	1	1	0	0
1	1	1	1	0	1	0	0

Symbols:

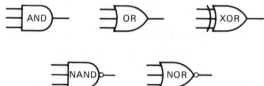

indicated by a small circle at the output of the gate symbol. A similar circle at an input indicates negation of that particular input. It is customary to omit the lettering from the symbols in technical work.

A convenient notation for inversion of a logical state A is $\overline{A}$, read as "A-bar," "not-A," or " the complement of A." The AND operation is denoted by a dot, so that A · B means A and B; whereas for the OR operation the plus sign is used. The exclusive-OR is denoted by an encircled plus, $\oplus$. Thus, for example, the three-input OR gate performs the operation A + B + C, while a NAND gate gives the output $\overline{A \cdot B \cdot C}$. This symbolism is adopted from Boolean algebra.

The *exclusive-OR* is a logic element that corresponds more closely than the OR gate to the everyday use of the word "or." The output is 1 if A = 1, *or* if B = 1, but *not* if both inputs are 1. The exclusive-OR can be assembled from a combination of AND and OR gates with the addition of inverters, as shown in Figure 8-8. For simplicity, only two inputs are shown. If there

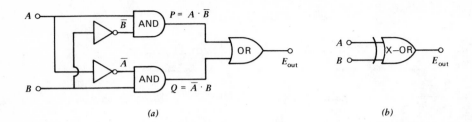

(a)   *(b)*

Figure 8-8. EXCLUSIVE-OR gate with two inputs: (a) showing internal logic, (b) symbol, and (c) truth table.

are x inputs, then x AND-gates of x inputs each, as
well as x inverters, must be supplied.

TABLE 8-2

Truth Table for Example in Text

A	B	C	D	Out
0	0	0	0	0
0	0	0	1	0
0	0	1	0	0
0	0	1	1	0
0	1	0	0	0
0	1	0	1	0
0	1	1	0	1
0	1	1	1	0
1	0	0	0	0
1	0	0	1	0
1	0	1	0	1
1	0	1	1	0
1	1	0	0	0
1	1	0	1	0
1	1	1	0	1
1	1	1	1	0

SYNTHESIS OF GATE SYSTEMS

In practice, it is often necessary to assemble logic systems to respond in specified ways to specific combinations of input states. In an industrial system, for example, we might want to actuate a given mechanism only for specific combinations of sensor signals. Thus, in a system of four sensors, one might want to actuate a load corresponding to the combinations shown in Table 8-2.

To implement such a table, it is always possible to use a combination of AND and OR gates, simply OR-ing all the cases in which the output is 1 (Figure 8-9). It is often possible to simplify the circuitry compared to the OR approach. Examination of the truth table for this example will show that the logic can be expressed as "(A or B) and C and $\overline{D}$." This leads to the diagram of Figure 8-10.

A rather general example of simplification is shown in Figure 8-11. Note that the circuit in Figure 8-11a is acting as a NOR gate, whereas that in Figure 8-11b is doing a NAND job. This equivalence between OR and AND functions is quite useful when one does not

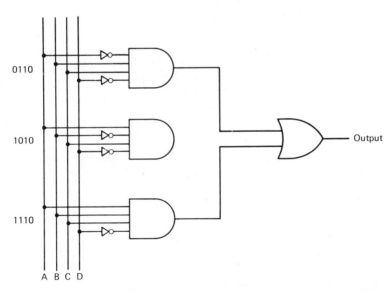

Figure 8-9. A "brute-force" implementation of Table 8-2.

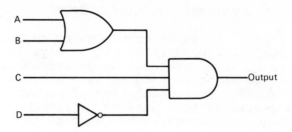

Figure 8-10.   Simplified logic circuit implementing Table 8-2.

have a given gate available and must synthesize it.
This combination rule parallels a well-known logic formula called the De Morgan theorem:

$$\overline{A} \text{ and } \overline{B} \;=\; \overline{A \text{ or } B}$$

$$\overline{A} \text{ or } \overline{B} \;=\; \overline{A \text{ and } B}$$

Note that in accordance with the black-box concept, only the truth table is operationally significant, not the details of assembly. The equals signs in
the De Morgan rules have just this meaning.

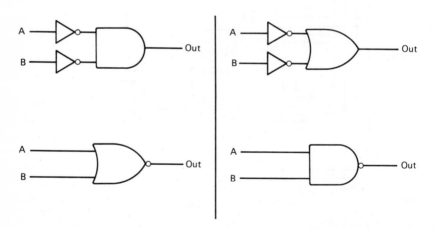

Figure 8-11.   Combinations of gates to illustrate equivalence of
functions.

CONSTRUCTION OF GATES

The major requirements of a logic gate in terms of performance are that it must have (1) low input current, (2) high output current capability, (3) low heat dissipation, (4) high speed of transition, and (5) high noise immunity. Historically, digital integrated circuits have passed through several stages of development, and a variety of families of gates are now available. The most popular of these are the so-called *transistor-transistor logic* (*TTL*) and the *complementary-metal-oxide-semiconductor* (*CMOS*) types.

The TTL units are available in a large variety of standardized forms (the "7400 series"). They are low priced and robust, and their rather high input impedance guarantees that one gate can feed signal to as many as 10 inputs of other gates in parallel (*fan-out* of 10). The transition time is approximately 10 to 20 nsec, which is sufficiently fast for most laboratory applications. The power dissipation (about 20 mW/gate) could add up to a rather high value in a large instrument.

Of considerably lower power dissipation (1 mW) are the CMOS gates. They permit the use of smaller power supplies in extensive systems, and lower the worry about heat build-up. More important yet, for a given maximum dissipation *per package*, one can have 20 times as many gates as compared to TTL. For laboratory use, the major advantages of CMOS are the very low price per gate, and the possibility of using a variety of power supplies (3 to 15 V). On the other hand, CMOS tends to be slow (250 nsec) and prone to failure caused by electrostatic charges. (There is little danger of this when the gates are in circuit; the damage can occur when the IC is handled out of the circuit.)

As mentioned before, the 7400 series is the most widely used line of TTL components, and it is manufactured by many firms. The last digits of the designation specify individual devices in the series. Some of these are listed for reference in Table 8-3. All are packaged in "dual-inline packages" (DIPs) with 14 or 16 pins.

The description "quad 2-input" means that the device contains four identical gates, each with two inputs. It is recommended that an experimenter choose as general-purpose units 3-input gates rather than

TABLE 8-3

Selected 7400-Series TTL Gates

Function	Type number
Quad 2-input NAND gate	7400
Triple 3-input NAND gate	7410
Triple 3-input AND gate	7411
Quad 2-input OR gate	7432
Quad 2-input NOR gate	7402
Triple 3-input NOR gate	7427
Quad 2-input Exclusive-OR	7486
8-input NAND gate	7430
Hex inverter	7404

those with only two inputs, as they are considerably
more versatile. Unused inputs can be tied to active
ones without any change in logic.

Most of these devices have counterparts that are
designated as "open collector" types. In these, the
transistors in the output circuits are not connected
internally to the power supply, and so such a connection must be made externally. This permits use of a
higher voltage to drive some device such as a CMOS
gate or a small relay for which 5 V may not be sufficient.

LOGIC GATES IN PATTERN RECOGNITION

As indicated previously, the NOR gate is able to
single out the 000 input, whereas the AND gate is able
to identify the 111 state. In other words, of all
possible combinations, the gate responds only to a specific pattern of ones and zeroes. This capability is
even more enhanced when combinations of gates and inverters are used to identify and single out a given
input combination. Consider, for example, the ASCII
symbol "carriage return," which is identified by the
pattern 1000001101. To identify this symbol, we need
to have a system that coldly ignores any pattern except the carriage return command at which action is to
be taken. The circuit in Figure 8-12 shows how such
an identification can be made.

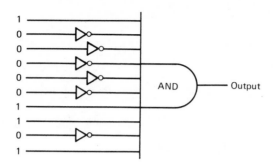

Figure 8-12. A ten-input AND-gate used to recognize the binary word 1000001101. The output is at logic "1" if this sequence of ones and zeroes is present at the inputs and not otherwise.

BINARY NOTATION

The familiar decimal system of numeration is not convenient as a vehicle for expressing digital information, where only two levels are present rather than ten. For this purpose, one can use a binary representation of numbers, the *binary system*. It can be visualized by reference to Table 8-4. The similar structures of the binary and decimal systems can be better understood by observing that in both systems, as we increment a number by 1, the number in the right-hand column ("least significant bit," LSB) is increased by 1 until we run out of digits, namely at 9 (decimal) or 1 (binary). In each case the next increment returns the LSB to 0 and carries 1 to the next place to the left. There are other similar systems, octal (using 0 to 7) and hexadecimal (using 0 to 9 and ABCDEF), both widely used in computer technology.

ENCODING AND DECODING

The logic gates described previously have in common the characteristic that they generate a single logic line out of several inputs. This process can be generalized. For example, in Figure 8-13 is shown a logic unit performing a 4-to-2 line conversion. The circuit assumes that one and only one of the four inputs is at logical one at any given time. This would

TABLE 8-4

Comparison of Binary and Decimal
Enumeration

Decimal number	Binary number
00	0000
01	0001
02	0010
03	0011
04	0100
05	0101
06	0110
07	0111
08	1000
09	1001
10	1010
11	1011
12	1100
13	1101
14	1110
15	1111

occur, for instance, in an experiment where the position of an animal is monitored by various sensors in
his cage. Clearly the animal is only at one place at
a time. The circuit is an example of an *encoder*, and
can represent a considerable saving in the number of
signal lines required. Note that the number of input
lines is equal to the maximum number of combinations
that can be represented by the output. Thus three out-

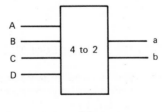

(a)

Figure 8-13. A 4-to-2 encoder and its truth table. Note that
only one input is permitted to be high at any given moment.

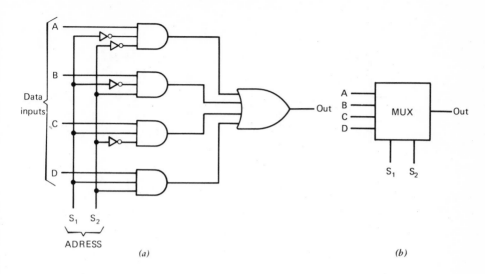

Figure 8-14.   (a) A four-line multiplexer;   (b) its symbol (MUX
is a common abbreviation):   (c) the selection table. The four
possible combinations of S_1 and S_2 select which of the four inputs
will determine the output. The states of the other inputs is immaterial. The "address" and " data" inputs correspond to usual
computer parlance.

put lines can encode up to binary 111, decimal 7, which,
including 000, gives eight possibilities. Similarly,
four lines can encode 16 inputs, etc.
     The reader should realize that such an encoding,
by necessity, must ignore some of the possible input
combinations. For example, 8 input lines can have as
many as 256 combinations, of which the decoder can only
single out 8, namely those composed of a single 1 and
seven 0's.
     A rather special type of encoder, the *multiplexer*,
selects which of a number of logic lines determines the
output state (Figure 8-14). Let us assume that S_1 is
high and S_2 is low (i.e., address code 1, 0). Then the
inputs to the gate a are (A, 0, 1) which can never
give a high output since one of the inputs is zero.
The reader can easily ascertain that only gate c can
have a high output. The OR-gate will have as inputs
(0, 0, C, 0). Thus code (1,0) selects input C to be
transmitted to the ultimate output. The circuit be-

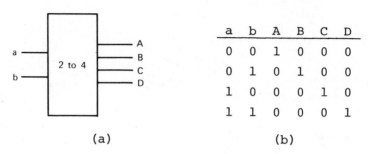

a	b	A	B	C	D
0	0	1	0	0	0
0	1	0	1	0	0
1	0	0	0	1	0
1	1	0	0	0	1

(a)   (b)

Figure 8-15.   (a) The 2-to-4 decoder, and (b) its truth table.

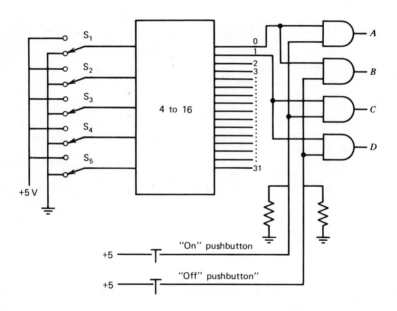

Figure 8-16.   An industrial control for a bank of 32 valves. The manual selector selects which line is to be addressed. No action is taken until one of the pushbuttons is pressed, when the designated motor is actuated. Note that a computer could operate the whole system by means of seven control lines.

haves as if it consisted of a 4-position mechanical switch capable of operating within a fraction of a microsecond.

The multiplexer is available as one IC unit and is useful whenever one wants to monitor several lines one at a time. To come back to the example of an animal in a cage, let us assume that the right and left halves of the cage are electrically rigged to indicate the presence of the animal at the right by logic 1 and on the left by logic 0. A collection of cages can then be connected to a multiplexer, so that any particular cage can be selected by an appropriate code presented to the address inputs, S_1 and S_2.

The operation of *decoding* is the reverse of encoding; a few input lines generate many output lines. An example is shown in Figure 8-15, where two inputs select which one of four outputs is high. Such decoders are useful in remote control, since the controller can have fewer outputs than the number of lines to be controlled. An example of such an application is shown in Figure 8-16. The remote control station has five selector switches to choose the code for up to 32 valves. Thus valve #0 is coded (00000), valve #1 has the code (00001), and so on. No action is taken until either the "ON" or "OFF" button is depressed, at which time a servo mechanism closes or opens the selected

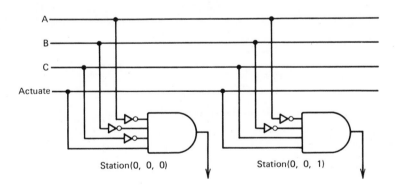

Figure 8-17. Example of remote control by means of a bus. The decoders are activated only if the proper combination of signals (address) appears on lines A, B, and C, and if the ACTUATE line is high.

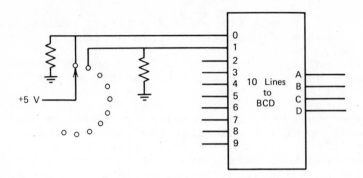

Figure 8-18. A 10-line to BCD encoder. This particular variety of BCD code is called the 8-4-2-1, from the decimal values of the bits ABCD. Other codes are also available.

valve. The valves might be at inaccessible or hazard-out places, or it might simply be desirable to control them all from a central location.

Considerably more flexible is the address decoder, where the controller is connected to a single set of lines (called a *bus*) to which various devices are connected in parallel, as shown in Figure 8-17. The advantages of such bus connections are that one can connect and disconnect any device at any time, and one can change physical locations simply by moving into a convenient spot on the bus. The bus in Figure 8-17 can only accept eight control points (binary 000 through 111), called addresses. A larger number of

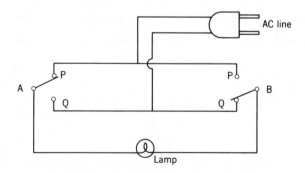

Figure 8-19. A pair of three-way switches to control a light.

lines in frequently encountered in practice. Thus 10
lines, plus Actuate, can control 2^{10} = 1024 addresses.
      So far we have assumed that the input (control)
lines are "binary" in the sense that the combinations
have to be reckoned as binary numbers. For the sake
of operating simplicity, it is more convenient to select lines by means of decimal numbers. Integrated
circuits are available to make the transformation. Usually an intermediate code is generated, the so-called
binary coded decimal (BCD), which is a representation
of a decimal number as a 4-bit binary equivalent that
is allowed to take only 10 values. An example is
shown in Figure 8-18, where a switch with 10 positions
generates a binary code representing its position.
Note that binary numbers over 1001 are forbidden, since
they do not make sense in this scheme. The BCD code is
frequently used in digital voltmeters and printers,
with a group of four digital lines assigned to each
place in the decimal number.
      It will be instructive to consider a number of
everyday situations in which logic decisions have to
be made. For example, Figure 8-19 shows a so-called
"3-way switch," permitting on-off control of a light
from two locations. It is left to the reader to establish that this operates as an exclusive-OR gate. Somewhat similarly, a 10-floor elevator control can receive orders from many floors and encode them to give
requisite floor commands. A household burglar alarm
system can also be considered as a combination of logic
gates: if window switches A, B, C, etc., are all closed, the alarm is at standby, but if any one switch
opens, the alarm sounds.
      A much more complex switching system, too elabor-
ate to consider in any detail, is the telephone system.
Each 7- or 10-digit number is encoded by action of
the dial or pushbutton equivalent. The encoded information is then transmitted by wire and wireless, to be
decoded at the remote location so as to ring the correct station, out of the millions possible.
      Another example of encoding is found in some var-
ieties of electric typewriters, which translate the
pressing of a key into a binary code. This is decoded
to provide commands to control the position of the
typing element. It is of interest to note that too
light a touch on any key may result in a zero code symbol, which in one brand, corresponds to a hyphen.

Yet another application of gating is the coincidence-anticoincidence circuit, important for very low-level radioactivity measurements, where cosmic radiation would interfere (Figure 8-20). This system contains an array of radiation counters around the measurment area. Whenever a photon of external radiation (such as a cosmic ray) penetrates into the measurement area, one or more of the counters of the array produces a signal. After level conversion to 5 V (not shown) these signals are combined into a 6-input OR-gate, which gives a logical high at point X whenever a cosmic ray penetrates the system. In the logic setup shown, the AND gate will produce an output pulse only for the case "A and B but not X." This means that both A and B counters have sensed radiation, but the outside ring counters have not. Cosmic rays have a much greater penetrating power than the inside source radiation that is being measured. The latter can never get as far as the outer ring of counters because of the shielding interposed. Hence the required logical condition can only be fulfilled by inside radiation, thus removing the background due to cosmic rays almost completely. This amounts to introducing an impenetrable electronic shield around the experiment.

PROBLEMS

8-A. Design a gate circuit to implement the table:

A	B	C	Out
0	0	0	0
0	0	1	0
0	1	0	1
0	1	1	0
1	0	0	1
1	0	1	0
1	1	0	1
1	1	1	0

8-B. Write the truth table for the circuit of Figure 8-21.

8-C. Design a circuit to perform the operations indicated in the table. Use the minimum number of gates.

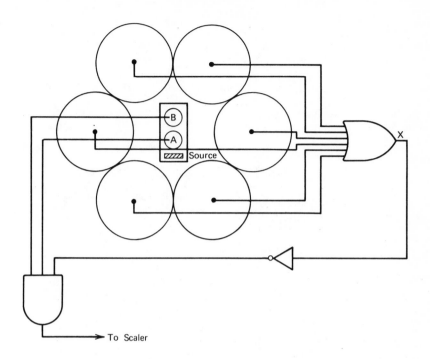

Figure 8-20. Coincidence-anticoincidence circuit. Necessary
high-voltage circuitry not shown. The signals from A and B are
said to be in coincidence, and they are in anticoincidence with
signals from the outer ring of counter tubes.

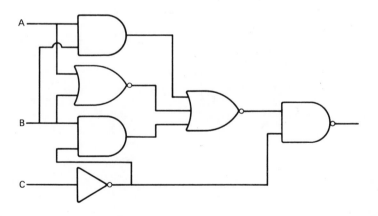

Figure 8-21. See Problem 8-B.

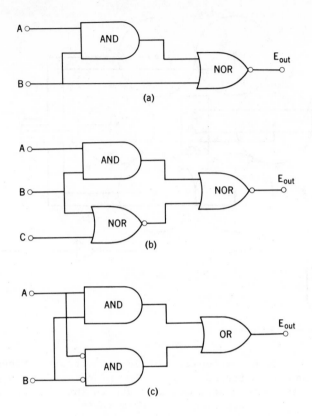

Figure 8-22.   See Problem 8-1.

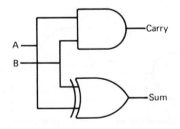

Figure 8-23.   See Problem 8-3.

A	B	C	Out_1	Out_2
0	0	0	0	1
0	0	1	1	0
0	1	0	1	0
0	1	1	0	1
1	0	0	1	0
1	0	1	1	0
1	1	0	1	0
1	1	1	0	0

8-D. Implement the function (A or B) and $\overline{(C\ or\ D)}$

8-E. What are the decimal equivalents of the following binary numbers?

          1101101
          0101000
             1110
          0101010

8-F. What are the binary equivalents of the following?

          39,   65,   119,   250

                    *   *   *

8-1. Establish truth tables for the circuits of Figure 8-22.

8-2. Devise logic circuits that have the following
truth tables:

A	B	E_{out}
0	0	0
0	1	0
1	0	1
1	1	0

(a)

A	B	E_{out}
0	0	1
0	1	0
1	0	0
1	1	0

(b)

A	B	C	E_{out}
0	0	0	0
0	0	1	1
0	1	0	0
0	1	1	1
1	0	0	0
1	0	1	1
1	1	0	0
1	1	1	0

(c)

8-3. Give the truth table for Figure 8-23, and explain why the two outputs are called "sum" and "carry."

8-4. Consider the following table showing correspondance between decimal numbers and a special digital code called the *Gray code*:

Decimal No.	Gray Code			
	A	B	C	D
0	0	0	0	0
1	0	0	0	1
2	0	0	1	1
3	0	0	1	0
4	0	1	1	0
5	0	1	1	1
6	0	1	0	1
7	0	1	0	0
8	1	1	0	0
9	1	1	0	1

Design a logic system to convert Gray code to BCD.

8-5. The circuit of Figure 8-24 can be used to generate 4 lines out of 3. Write its truth table.

8-6. Construct a 16-input NAND gate using 7420 four-input NAND gates and 7404 inverters if needed.

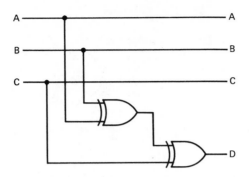

Figure 8-24.   See Problem 8-5.

8-7.   What is the decimal equivalent of each of the
        following binary numbers?

        (a) 01001,   (b) 11000,   (c) 10001   (d) 11111

        Write binary equivalents of the following:

        (e) 17,   (f) 168,   (g) 4,   (h) 44.

IX

FLIP-FLOPS AND COUNTERS

Up to this point we have considered logic systems in which there is a continuous correlation between input and output. Any change in the input states is reflected practically instantaneously by a corresponding change in the output. However, there are many situations in which it is necessary to retain the state of logic present at some previous moment. To accomplish this, logical elements with *memory* are required. The concept of memory is quite distinct from that of time delay, which may account for an output holding its state for a short time after a change of input signals. True memory devices may be called on to retain their information for long periods. Indeed, some kinds hold their logic state even when the power is turned off.

Digital computers require very extensive memory capability and usually employ some type of magnetic component for the purpose. This is because magnetic materials can retain magnetization in either of two possible states corresponding to orientation of the field in two opposite directions. These memory units can be fabricated in many physical forms, including magnetic tape, discs, and cores. The details are outside the scope of this book.

FLIP-FLOPS

For most laboratory electronic instruments, where memory is required, it is best implemented with *flip-*

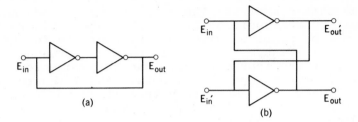

Figure 9-1. Basic flip-flop made from two inverters connected in tandem. (a) The simplest representation; (b) the same circuit arranged to demonstrate its symmetry and provided with two inputs and outputs.

flops. A simple form is shown in Figure 9-1a. In the absence of any connection at the input, the output can be either "1" or "0," maintained by feedback through two inverters. A momentary contact of the E_{in} terminal to either logic level will bring the output to the same level, and the feedback will keep it there, even when the contact is removed. This amounts to a bistable memory device, equally able to retain either logic level.

Figure 9-1b shows the same circuit redrawn to emphasize the equivalence of the two inverters. A second input and output have been added to further em-

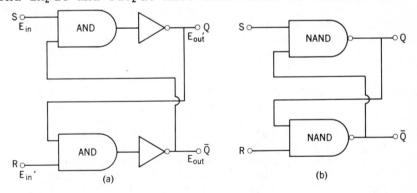

Figure 9-2. Flip-flop made from NAND-gates: (a) stressing the relation to Figure 9-1; (b) a more compact representation; (c) the truth table. The conventional notation is here adopted: R and S represent *reset* and *set*, respectively; $\overline{Q}$ is the inverse of Q, except when R and S are both zero, in which case $Q = \overline{Q} = 1$.

phasize this symmetry. The circuit can be made more
flexible by the addition of two AND gates (Figure 9-2a),
which can be combined with the inverters to form NAND
gates, as in Figure 9-2b. This permits permanent connections at the inputs rather than only momentary contacts. In normal operation, both R and S inputs are
kept at logic "1." Momentary transition to "0" at the
S input produces $Q = 1$ and $\overline{Q} = 0$, whereas a momentary
grounding at R produces $Q = 0$ and $\overline{Q} = 1$. Both states
are self-sustaining. They are called *set* and *reset*,
respectively (set to "1" and reset to "0"). After the
flip-flop is set, subsequent changes of S are immaterial. The same holds true for the reset mode. If R and
S are both made "0", then Q and $\overline{Q}$ will both become "1";
on returning both inputs to "1", the outputs will un-
predictably go to either set or reset.

This type of flip-flop, designated RS, can be
used to remember whether a momentary grounding (pulse
of logic "0") has or has not occured at the S input.
Assume that the circuit is initially in its normal
state with both R and S equal to "1". The state of Q
answers the question whether or not S has been grounded.
If $Q = 0$, this may be interpreted as the answer "no,
it has not." A grounding pulse at S changes Q to "1,"
meaning "yes, it has." The answer is still "yes" after
additional groundings at S, but a grounding pulse at R
resets the circuit to $Q = 0$. One can alternatively
think of the circuit as indicating *which* input has most
recently received a pulse of "0," the set indicated by
$Q = 1$, or the reset by $\overline{Q} = 1$.

Figure 9-3 shows a somewhat more complicated circuit, the *gated* or *clocked RS* flip-flop, in which two
NAND gates are inserted at the inputs. If the so-
called clock input C is at "1," the NAND gates act simply as inverters. Consequently the functioning is the
same, but now the circuit remembers whether a pulse to
"1" rather than to "0" has occurred at the input.

When the clock signal drops to "0," the mode of
operation changes, since both input NAND gates are now
forced to "1" regardless of R and S. The output is
thus frozen in whatever state it was in just before
the change in the level at C. The two modes of operation can be considered as *sample* and *hold*. If during
the sampling period ($C = 1$) S was at "1" and R was "0,"
the output Q becomes and remains "1." If on the other
hand, R was "1" and S was "0," then Q will be "0." If

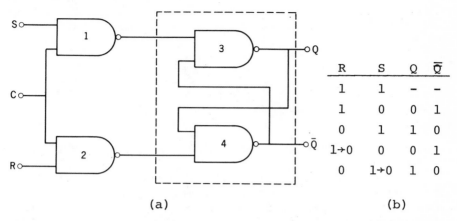

R	S	Q	$\overline{Q}$
1	1	−	−
1	0	0	1
0	1	1	0
1→0	0	0	1
0	1→0	1	0

(a)   (b)

Figure 9-3. Clocked *RS* flip-flop. (a) Schematic; the part in the
dashed box is a basic *RS* unit; input NAND-gates have been added;
C is the clock input. (b) Truth table. Observe that this table
is exactly opposite to that in Figure 9-2. The clock is assumed
to be initially in its "1" state. When the clock changes to zero,
the previous state is conserved, except for the first case, which
is undefined.

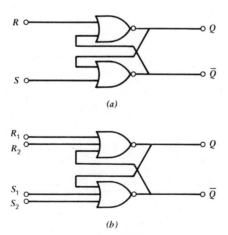

Figure 9-4. An *RS* flip-flop made from NOR-gates: (a) conventional
form, (b) modified to provide multiple inputs.

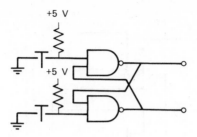

Figure 9-5. An *RS* flip-flop acting as a bounceless switch.

both inputs were at "0" during the sampling, the state
of the unit will remain the same as before the clock-
ing cycle.

Both R and S should not be made "1" simultaneously. In this case, during sampling, both input NAND
gates have outputs at "0,", since R, S, and C are all
"1." Both gates change to "1" when the clock signal

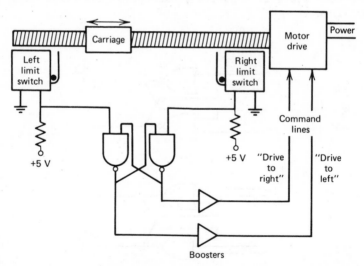

Figure 9-6. Limit control system for an automatic scanning mechanism. The carriage when driven to the left eventually closes the
"left limit switch," causing the motor to reverse and driving the
carriage to the right, when it will close the "right limit switch,"
reversing again, and so on.

goes to "0." It is, however, unavoidable that the gate
return to "1" at slightly different times, with the result that either $Q = 0$ or $Q = 1$, depending on which of
the two gates was the first to change state.

Flip-flops can also be made with NOR-gates, as
shown in Figure 9-4. The normal state of the inputs
is, in this case, "0."

RS flip-flops are useful in interfacing mechanical and electronic logic devices. An example of such
use for a switch is indicated in Figure 9-5. A purely
mechanical switch would probably suffer from "bounce,"
a series of fast on-off transitions on switching that
result from actual mechanical bouncing. This can ad-
versely influence logic systems. Since both R and S,
once activated, are no longer affected by changes in
their respective lines, such bouncing is ignored after
the first contact. In other words, the gate *latches*
on to its state.

Another way in which latching can be used is in
automatic scanning devices such as spectrophotometers
and zone refiners. The heart of such a set up is an
RS flip-flop (Figure 9-6), which could be either electronic or its mechanical counterpart, a latching relay.

THE D FLIP-FLOP

A characteristic of the RS flip-flop when used
as a memory is that two input lines are needed to control the unit, one for setting to "1," the other for
resetting to "0." A more versatile arrangement is exhibited by the D flip-flop as shown in Figure 9-7.
This also uses two lines, but now only one of them de-

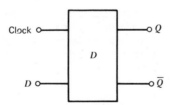

Figure 9-7. The D flip-flop. The D input is transferred to Q
whenever the clock changes from "0" to "1." This is called
leading-edge triggering.

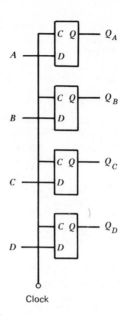

Clock

Figure 9-8. A four-bit register composed of D flip-flops. Each
of the input lines ABCD can be considered as carrying one digit
of a 4-place binary number. The register will accept this number
on receiving a 0-to-1 clock transition and retain it until the
next clock pulse.

termines the logic state, whereas the other, the clock
line, is used to determine the exact moment when this
logical state is to be loaded into the memory.
    The major advantage of clocked flip-flops is that
many units can be made to load their particular input
values simultaneously. This is called *strobing* or
synchronizing.
    A group of D units connected together and used to
retain the states of a number of lines simultaneously,
forms a D *register*, an example of which is shown in
Figure 9-8. Registers are widely used in computers
and are very useful for retaining various logical data
for future reference. For example, one might want to
store the instantaneous values of various logic lines
that change too fast for direct observation. For this
purpose one might use the circuit shown in Figure 9-9.
The operation is similar to the HOLD operation in analog electronics, when one wants to examine a fast
moving process at leisure. Assume, for example, in

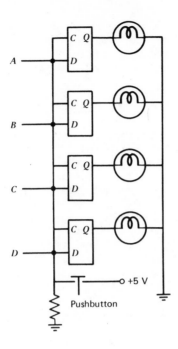

Figure 9-9. A register used to "freeze" four-digit variables by pushbutton command. In practice it might be more power efficient to connect the lamps between $\overline{Q}$ and +5 V; the logical result is the same.

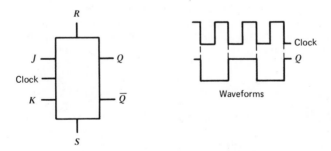

Figure 9-10. An example of a *JK* flip-flop. The connections *J*, *K*, *R*, and *S* are, in this application, all at logic "1." Other models of *JK* flip-flops vary in details of truth table and wave forms.

Figure 9-9, that at the moment one presses the button, A, B, C, and D have the values 1, 0, 0, and 1, respectively. Then regardless of any change in logic levels, lamps 1 and 4 will light and remain lit.

THE *JK* FLIP-FLOP

An even more versatile component is the *JK master-slave* flip-flop (Figure 9-10), which actually contains two flip-flops in one unit. In its fully extended form, it has no less than seven connections to external circuitry, not counting the power supply. If all inputs, R, S, J, and K, are at logic "1," the mode of operation is such that the flip-flop interchanges states whenever a 1-to-0 clock pulse is applied (*toggling* or *complementing*). The state remains the same when the clock returns to high. Toggling can be inhibited by logic signals at J or K. The circuit can be set to $Q = 1$ independently by grounding the S input, or reset ($Q = 0$) by grounding R. These commands override any signals present at other inputs, including the clock.

The presence of the clock input in the *JK* flip-flop is of great inportance in that it ensures lack of any interference between an incoming signal and the information previously stored in the flip-flop. This is a consequence of the fact that the unit contains two cascaded *RS* flip-flops, the master and the slave. Input information is fed into the master flip-flop only when the clock signal goes high; in this state there is no communication between master and slave. The information is passed to the slave as the clock goes low. The result of this sequence is that data from both the present and prior cycles are stored simultaneously.

MONOSTABLE AND ASTABLE FLIP-FLOPS

The flip-flops described above are inherently bistable devices; they will remain in either of two stable states indefinitely. Similar circuits with other modes of operation are also possible. For example, some types have internal connections that ensure that a change of state is automatically followed by a return to the original condition. The resulting cir-

cuit is *astable*; it requires no input and becomes a
type of oscillator producing a continuous train of
square waves. This is usually called a *multivibrator*.
On the other hand, if the feedback action is only one-
sided, a *monostable* circuit results. A change of input
generates an output pulse, followed by resetting of the
circuit to the original condition. The monostable is
also called a *one-shot* multivibrator or a *univibrator*.

Any of these flip-flops can be assembled from
discrete components, but they are usually fabricated
in quantity as complete integrated circuits at considerably lower cost. A widely used TTL type of monostable is the 74121, which can give pulses varying
from 40 nsec to 40 sec following an input transition.
The adjustment of the pulse width, τ, is established
by an external resistor and capacitor, following the
formula $\tau = 0.7\ RC$. Among square-wave generators, the
type 555 mentioned in Chapter 7, is very widely used.
It can supply monostable operation as well as astable.

A monostable can be made to alter the width of
pulses, as shown in Figure 9-11. This is useful because it changes the duration of the operation control-

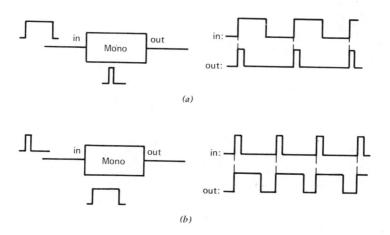

(a)

(b)

Figure 9-11. Examples of (a) pulse shortening, and (b) pulse
lengthening. The monostable in this case responds to a positive-
going logic transition.

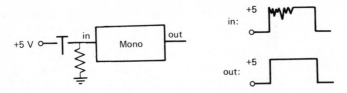

Figure 9-12. A momentary contact pushbutton switch rendered bounceless by means of a monostable.

led by each pulse without changing the repetition rate; for instance, for lengthening the duration of a visual signal, or for controlling the power delivered to a load (called "duty-cycle variation").

It was mentioned earlier that an *RS* flip-flop can be used to debounce on-off switch operations. An analogous situation for momentary contact switches can be cured by the use of the circuit of Figure 9-12. This prevents multiple toggling of a *JK* flip-flop controlled by the switch provided only that the pulse width of the monostable is longer than the bounce time.

For applications requiring the timing of a series of events with a precision of 1% or so, a chain of monostables as in Figure 9-13 could represent a very convenient and inexpensive solution. In this system,

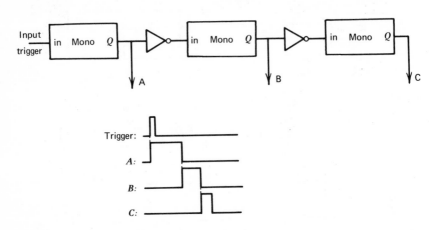

Figure 9-13. A sequential controller using a string of monostables and initiated by a trigger signal, for example a pushbutton. Since many monostables offer complementary outputs , Q and $\bar{Q}$, the inverters may not be necessary. Some monostables are triggered on $0 \rightarrow 1$ and some on $1 \rightarrow 0$ transitions.

each monostable after the first is triggered by the
trailing edge (1 to 0 transition) of the pulse of the
preceding unit.   The applications to automation are li-
mitless, especially since one can form branches and
loops of any desired complexity.   Higher timing precision, however, can be obtained by means of shift registers and counters, to be considered in the next section.

SHIFT REGISTERS

The nature of the flip-flop is such that it re-
cords one logical value or one *bit*, whereas groups of
flip-flops can memorize groups of logical values
(words).   Thus a group of six flip-flops can record a
6-bit word.   There are other devices available that,
in addition to logic storage, can provide logic manipulation.   The simplest form of such manipulation is
shown in the shift register of Figure 9-14.   The four
inputs, A, B, C, and D, can load any given pattern of
ones and zeros, when the preset enable is held at logic "1."   After loading, the clock transitions cause a
shifting of the information to the right.   The first
register value is transferred to the second, the second to the third, etc., until the last, whose information simply exits.   The first register loads the value
present at IN in the moment of the clock transition.
Note the interesting fact that connecting the

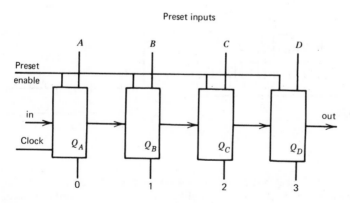

Figure 9-14.   A shift register.   The logical state at the input is
shifted through the system on   consecutive zero-to-one transitions
of the clock.

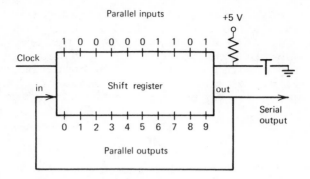

Figure 9-15. A shift register used to generate a repetitive pattern. Once loaded by presetting the parallel inputs, the pattern circulates through the register, and appears serially at the output.

output to the input avoids the loss of the information of the last flip-flop, since it is fed back into the first. This arrangement is called a *ring counter*, and can be used to generate a repeating pattern of ones and zeros. An example of such application can be seen in Figure 9-15, a device for generating the ASCII character "carriage return." The appropriate combination of binary digits is loaded in parallel into the register, and then by means of the clock is circulated through

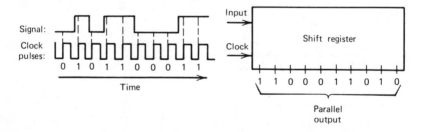

Figure 9-16. Serial entry into a shift register. Each clock pulse triggers the entry of a new value, while also shifting previously entered data to the right. The parallel output bits are in reverse order to entry because the first bit to enter moves the furthest to the right.

and appears in succession (serially) at the output.
The operation is called *parallel to serial conversion*.
   The opposite operation, *serial-to-parallel* conversion, is also possible, by simply entering the information into the input (Figure 9-16) by means of a succession of pulses, and reading it out at the outputs
labeled 0, 1, 2, 3, after the loading process is accomplished.
   Shift registers in ring counter configurations
are very useful for controlling a linear sequence of
operations. Typical of many experiments, one might
want to (1) take a measurement before the operation,
then (2) apply a stimulus, (3) wait for the response,
and (4) take a measurement of this response. Such a
scheme can be implemented elegantly by one shift register, as shown in Figure 9-17. The initial pattern,
1 000 000 000, ("read blank"), changes 1 sec later to
0 100 000 000 ("stimulus ON"). After 3 sec more, it
arrives at 0 000 100 000 ("read response"). After 6
sec more the cycle begins again.

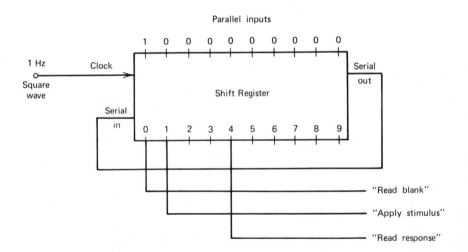

Figure 9-17. Shift register used to control a stimulus/response
cycle. The complete cycle is 10 sec in duration. The same basic
system can be used with most so-called pulse methods, which are
of the same stimulus/response type, though often faster. The unused outputs provide time for the "relaxation" of the system to
its original state.

The ring counter has practically no inherent time
limitation, so that cycles from microseconds to days
are equally possible. Note, however, that ring counters must be initialized every time the power is turned
on. This can be done by a combination of a CLEAR command, to eliminate all previous highs, and a PRESET
command to load the desired pattern.

COUNTERS

A register in which each flip-flop after the first
is driven by the output of the previous stage acts as
a *counter*. Figure 9-18 shows a four-bit combination.
All the J, K, R, and S inputs must be connected to
logic "1." The clock input, in its quiescent state,
is at level "0." When a "1" pulse comes along, it
changes the state of the master section of flip-flop
A, and on returning to "0" shifts the slave portion of
A to $Q = 1$, from its reset position at zero. Consequently one full cycle (0-1-0) at the input results in

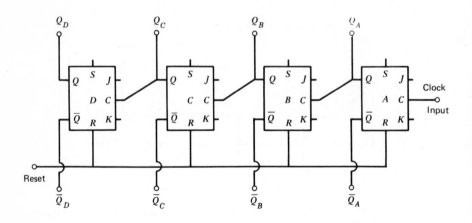

Figure 9-18. A 4-bit binary counter, composed of two dual JK
flip-flops, type 7476; the R terminals are at "1," except during
the reset operation; the S inputs are tied to +5V.

a half-cycle (0-1) at the output of A, which is com-
pleted by a second input pulse. Similarly four input
pulses are required to effect a cycle at Q_B, eight at
Q_C, and sixteen at Q_D. The sequence follows the table
of binary numbers from 0000 to 1111. Reset can be effected at any time by a "0" pulse on the R line.

A counter that does not somehow indicate its
state is useless. Its state can be read out if each
Q terminal is made to energize a lamp. Since flip-
flops have only two states, the numbers produced in
this way are in the binary system of numeration.

It is in general desirable to obtain a reading in
the familiar decimal system. The chief difficulty is
that the four flip-flops that are necessary to represent 10 are capable of counting to 16. Therefore some
means must be devised for stopping the counting at 10.
One way in which this can be done is to arrange logic
circuits to detect when binary 1010 is reached, since
this corresponds to 10 in the decimal system, and to
reset all flip-flops (Figure 9-19). For every power
of 10, four flip-flops and one resetting circuit are
required.

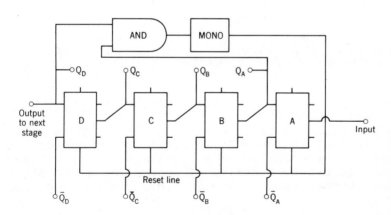

Figure 9-19. One stage of decimal counting, based on resetting
all flip-flops by an AND-gate, following the ninth pulse. The
monostable makes sure that the reset signal is held long enough
to clear all the stages. Note that the output sequence is in BCD
code.

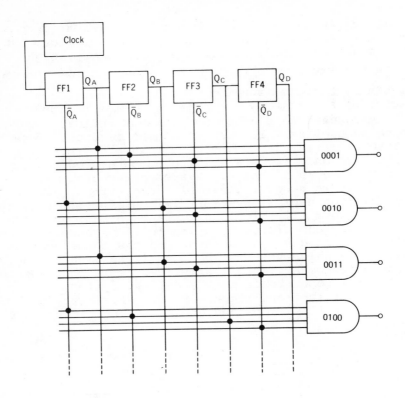

Figure 9-20. Timing circuitry showing the output gates corresponding to the first four combinations of Figure 9-21.

SEQUENTIAL TIMING

A counting register is useful in quite another way, in that it provides a flexible and convenient manner of programming events with greater precision than afforded by a string of monostables. Consider the binary number table shown in Figure 9-18. Each number can be uniquely identified by means of an AND gate. For example, number 3 (binary 0011) can be selected by a gate ANDing Q_A, Q_B, $\overline{Q}_C$, and $\overline{Q}_D$. Figure 9-20 shows circuitry to implement such a decoding procedure. Sixteen AND gates are required to complete the sequence. The actual wave forms so obtained are depicted in Figure 9-21.

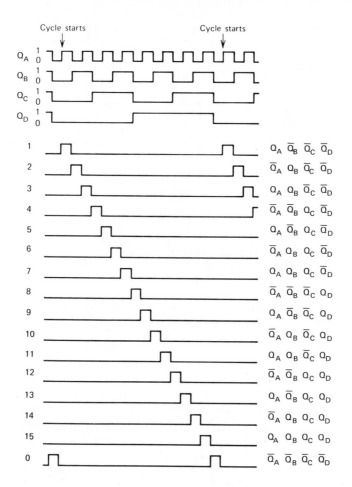

Figure 9-21. Timing sequence obtainable from the 4-bit flip-flop register of Figure 9-20, with a 1-Hz clock.

Let us suppose that the clock gives a square wave with a 1-sec period. The output of flip-flop A is a square wave with a period of 2 sec. It is important to note that each state, $Q = 1$ and $Q = 0$, lasts for 1 sec, which is the duration of the pulses at the output AND gates. The experiment to be sequenced is made to start at the same instant that the counter starts counting clock pulses. Then, if event X is to occur exactly 3 sec later and last for 5 sec, it can be ac-

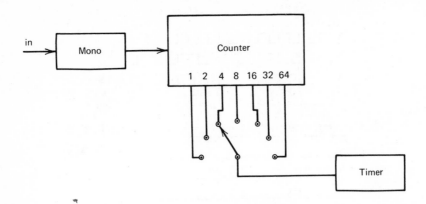

Figure 9-22. An example of a preset counter. After the number
of counts indicated by the switch position, the output of the
counter will go high, thus actuating the control that stops the
timer.

tuated by the signal from the AND gate marked 0011 and
turned off by 0101. Unless inhibited by logic, X will
repeat every succeeding 16 sec.

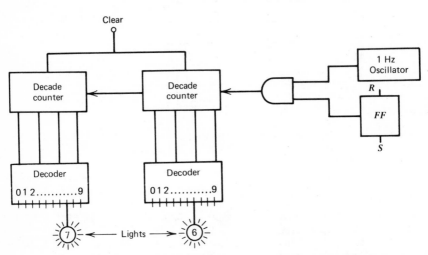

Figure 9-23. Timing circuit. Only one lamp from each decoder
can be ON at a given moment ("76" as illustrated), representing
times from 00 to 99 sec. Note the similarity between each decade
and the sequential controller of Figure 9-17.

Counters can also be used in a "preset" mode, meaning that they will give an indication when a preselect-
ed number of input pulses have arrived. The nuclear
counter described earlier could benefit by such a preset counter, as shown in Figure 9-22. Here the timer
measures the interval needed to reach a certain number
of counts. The timer itself can be implemented using
counters as shown in Figure 9-23. The 1-Hz oscillator
is gated on and off by an *RS* flip-flop which in turn
is manually set to start the counting. The system con-
tinues to count the number of 1-sec pulses until turned
off by a signal at *R*. The indicating lamps will con-
tinue to read the final seconds count until cleared.

Most timers and counters use direct digital readout rather than a string of lamps. This is made possible by the so-called *7-segment display* shown in Figure 9-24, widely encountered in digital calculators and
other equipment.

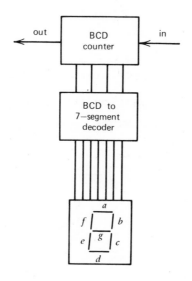

Figure 9-24. Example of a timer or counter driving a 7-segment
display. Various combinations of illuminated segments will give
any digit from 0 to 9. The decoder is rather complex; for example, it must energize segment *g* for the digits 2,3,4,5,6,8 and
9, while segment *d* is used for 2,3,5,6 and 8, and so on. Similar
displays are available for alphabetic symbols.

PROBLEMS

9-A. Devise a circuit to generate 1-Hz square waves, making use of the 60-Hz power line frequency, for application in the timing system of Figure 9-20.

9-B. Design a divide-by-12 counter.

9-C. Design a ring counter circuit able to generate continuously the SOS signal in Morse code.

9-D. Find out what a Möbius ring counter is and what it is good for.

9-E. In the circuit of Figure 9-20, a desired pulse starts 2 sec after time zero and finishes 10 sec later. Show how one can implement this sequence (a) by a "brute force" method using a multi-input OR-gate, (b) by the use of an RS-flip-flop and two decoding gates.

*   *   *

9-1. Construct a truth table for the circuit of Figure 9-25. Compare with Figure 9-3.

9-2. Describe the functioning of the circuit of Figure 9-26.

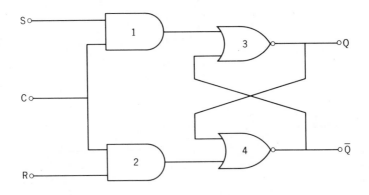

Figure 9-25. See Problem 9-1.

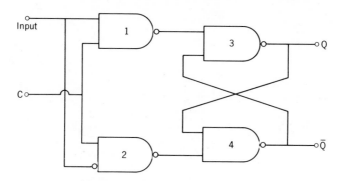

Figure 9-26.   See Problem 9-2.

9-3.   Describe the output waveform of the circuit of Figure 9-27.

9-4.   Draw the logic-gate equivalent of the direct-coupled transistor logic (DCTL) circuit of Figure 9-28.

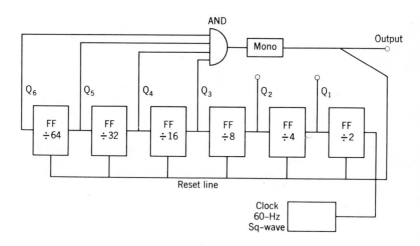

Figure 9-27.   See Problem 9-3.

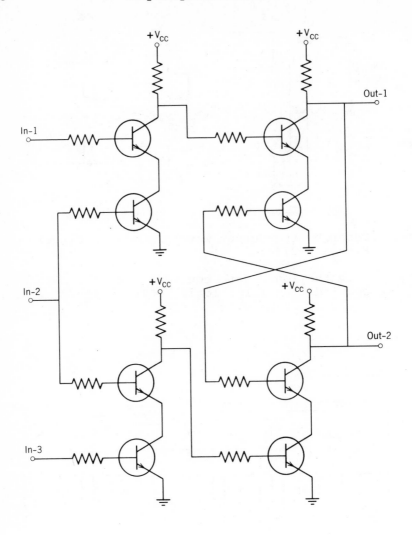

Figure 9-28.   See Problem 9-4.

X

INPUT AND OUTPUT DEVICES

To understand fully the design and operation of measuring systems, in addition to signal processing, one must also examine the devices that introduce information into the data processor (input transducers) and those that convert the data to a usable form (output devices). In computer parlance, the corresponding functions are grouped together as the "I/O" system (input-output).

TRANSDUCERS

Transducers may be classified as optical, mechanical, thermal, and so on, according to the energy domain from which information is to be converted into electrical signals. For each, one can write a "transducer equation," which relates the nonelectrical quantity of interest to the resulting output signal. For example, the glass electrode (at 25°C) gives an output that obeys the equation

$$E = K + 0.059(pH) \qquad (10\text{-}1)$$

where K is a constant. Such expressions can often be deduced from theoretical principles. In some instances, however, the theory of operation is inadequately developed to permit a mathematical treatment; the transducer equation must then be empirical, but not necessarily less useful on that account. Whether theoretical or empirical, the relation should be validated by examining the signals arising from systems with known properties. This is the procedure called *calibration*.

Transducers must also be considered with respect to their sensitivity and dynamic range, their frequency response, and the closeness with which they follow the working equation (often equivalent to *linearity*). Pre-valent sources of error must be delineated, including interferences from related sources, and from noise. There are frequently other limitations, such as maximum operating temperature and sensitivity to mechanical vibrations.

MECHANICAL TRANSDUCERS

The basic mechanical quantities to be observed are linear or angular displacements or deformations. The displacement may itself constitute the desired information, as in studies of thermal expansion, plant growth, and seismological motion of rocks. Alternatively, the displacement might be a secondary effect, for example the result of a change in fluid pressure. It might be periodic in nature, and one would speak of vibration studies or of sound waves.

It is often necessary to distinguish between measurement systems for small and large displacements, since the same transducers are seldom suitable for both extremes.

Small displacements can be measured by capacitive, inductive, or interferometric effects, all of which account for the motion to be observed through the effect on electromagnetic fields.

The capacitance between two parallel plates of area a cm^2, separated by a distance d cm in a medium of dielectric constant ε, is approximately $C = 0.0885$ $\varepsilon a/d$, where C is in picofarads. A change in capacitance can be effected by a variation of the separation d or by a change of the overlapping area a. The transducer equation is given, for instance, for a change of separation Δd, by

$$\Delta C = - k\Delta d \qquad (10-2)$$

where k is a constant.

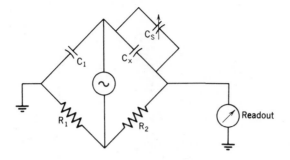

Figure 10-1. Comparison bridge for a capacitive transducer. The readout is a sensitive AC voltmeter or an oscilloscope.

Capacitance variation can be measured by means of a comparison bridge (Figure 10-1). At balance ($E = 0$)

$$C_S + C_X = \frac{R_2 C_1}{R_1} \qquad (10-3)$$

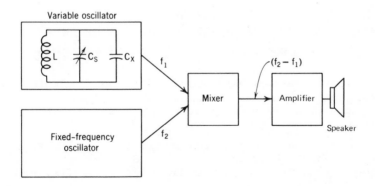

Figure 10-2. Heterodyne method for measuring a change in capacitance. The frequency $f_1 = 1/[2\pi\sqrt{L(C_S + C_X)}]$ can be adjusted by means of the calibrated variable capacitor, C_S, to equal the reference frequency, f_2. As the difference in frequencies decreases, the pitch of the sound produced by the speaker becomes lower.

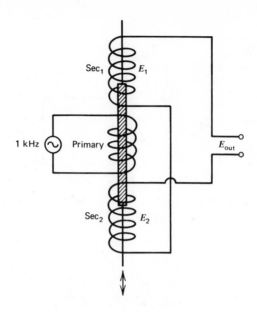

Figure 10-3. Linear variable differential transformer. The two secondaries are as nearly identical as possible, and connected in phase-opposition.

If the quantities on the right-hand side are held constant, variations in C_x can be offset by equal and opposite variations in C_s, a calibrated variable capacitor.

Another measuring circuit (Figure 10-2) involves a resonant LC tank in an oscillator. Again, changes in C_x are compensated by counteradjustment of C_s. The sensitivity of the system is such that the difference between the two frequencies noted in the figure can easily be made less than 20 Hz. Since the individual frequencies can be at least as high as 2 MHz, resolution to 1 part in 10^5 is possible. This can only be realized if care is taken in design and construction to make all reactances highly stable.

Another measuring device is a differential transformer (Figure 10-3). If the movable iron core is in its center position, the output $E_1 - E_2$, will be zero, but will vary linearly with displacement (usually up to

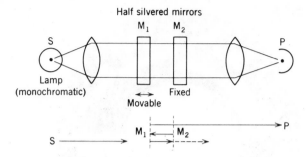

Figure 10-4. Fabry-Perot interferometer. Any relative movement
of the mirrors is observed as a sequence of interference "fringes."
Some of the light is reflected back and forth between mirrors, as
suggested in the lower diagram. The mirrors must be exactly parallel.

a few centimeters). The phase of the output changes by
180° in passing through null. If this transformer is
mounted vertically, so that the core hangs inside the
coils without contact, the resistance to motion can be
due entirely to magnetic forces that are proportional
to the displacement. This is just the condition needed for a precision analytical balance, and so a differential transformer can serve as a precise and convenient combined damper and transducer for recording bal-
ances.
   Small linear displacements can be measured with
the highest accuracy by optical interference. The moving part is made to carry a partially reflecting plane
mirror, which is part of an interferometer (Figure 10-
4). In this instrument, portions of the light beam
traverse different paths ($S-M_1-M_2-P$ compared to $S-M_1-$
$M_2-M_1-M_2-P$) to recombine at the photocell. If the path
difference is an integral number of wavelengths, the
two beams will be in phase, hence will reinforce each
other. Half way between such points, destructive interference will occur between the two out-of-phase
beams. The result is a series of *fringes*, alternate
light and dark periods, which cause similar fluctuations in the signal from the photocell as the movable
mirror is displaced. Since the light traverses the
increment of path twice, each fringe corresponds to a
displacement of one-half wavelength.

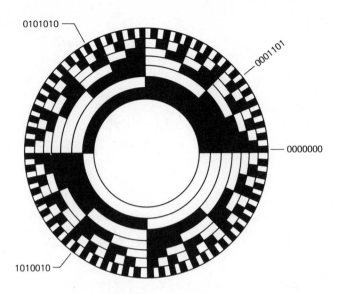

Figure 10-5. A seven-ring binary shaft encoder. The outer ring
has 2^7 = 128 segments. The binary numbers corresponding to a few
selected radii are shown. Photosensors at the position marked
"0000000" would see no light, and as the disc turns in the clockwise direction, would count up to 1111111, then repeat. Note that
if the disc is turning slowly, transitions may generate false
codes if not all photosensors switch at the same instant. To a-
void this difficulty, a different code can be used, namely the
Gray code where only one bit changes from one position to the next.

   With a conventional monochromatic light source,
such as the green radiation from a mercury arc (wave-
langth 0.546 μm), displacements are limited to a few
centimeters, but substitution of the coherent light
from a laser makes possible measurements up to much
greater distances. The resolution is limited to about
half the wavelength, a fraction of a micrometer.
   An important device for measuring angular displacements is the *digital shaft encoder*. This may take
the form of a disc (Figure 10-5) with concentric rings
that are divided into 2, 4, 8, 16, and so on, equal
segments, alternating black and white. The disc is
mounted on the shaft to be monitored. An array of tiny
photocells is so positioned that each looks at one ring,
responding to light reflected by a white area ("logic

1"). The resulting signals correspond directly to the
position of the shaft expressed as a binary number.
The output is easily handled by digital counting and
switching techniques.

Strain Gauges

The increase in resistance of a wire when stretched within its elastic limit forms the basis of a versatile device for measuring small displacements—the
resistance strain gauge. As a wire is stretched, it
increases in length and decreases in diameter, both
effects tending to raise the resistance. However, it
is generally found that the actual variation in resistance is different from that predicted by the geometrical alterations; the resistivity itself changes as a
result of the strain, and its coefficient can be of
either sign.
Strain gauges are characterized by a sensitivity
factor, S, defined as the ratio of the fractional change
in the resistance R to that of the length L producing it

$$S = \frac{\Delta R/R}{\Delta L/L} = \frac{L \Delta R}{R \Delta L} \tag{10-4}$$

For common materials S varies from about 0.3 for man-
ganin and 2.3 for nichrome to about 12 for pure nickel
and as high as 200 for some semiconductors.
Strain gauges are of two general types: bonded
and unbonded. Unbonded gauges are constructed in the
form of an array of fine parallel wires (Figure 10-6a)
clamped at each end to the two members whose separation is to be measured. The relative motion is limited to small displacements, of the order of 1% of the
length. Unbonded gauges are often used in pressure
transducers of the diaphragm type to detect small move-
ments of the diaphragm with respect to the instrument
case.
Bonded gauges are made of foil ribbon rather than
wire, and cemented onto a solid object which is subject to the motion to be observed (Figure 10-6b). This
type is particularly suited to measurement of torsion-
al or bending movement in bars or plates and is widely
used in engineering. It is also applicable to bal-
ances and load cells for measuring large masses and to
laboratory pressure gauges.

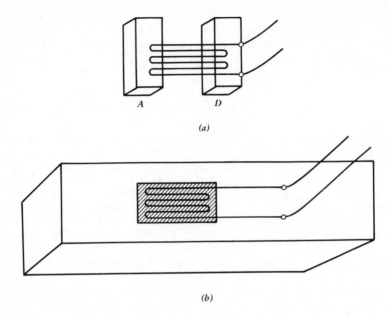

(a)

(b)

Figure 10-6. Strain Gauges. (a) Unbonded wire type; the folded wire is cemented or clamped at A and B, and responds to changes in the separation of these two areas. (b) an etched-foil (ribbon) gauge bonded to a metal bar to detect dimensional changes in the bar.

Strain gauges, being resistive devices, are often measured in a Wheatstone bridge circuit. It is convenient to use two gauges in a single assembly, one as the active element, the other for temperature compensation. Sensitivity can be doubled by using four gauges in the four arms of the bridge—two opposed ones for sensing strain, the others for compensation. If of the bonded type, the compensating gauges can be cemented to the same support as the active arms, but so oriented that they do not sense the motion, while nevertheless experiencing the same ambient temperature.

As in any DC bridge circuit, the power supply and amplifier cannot *both* be referenced to ground. It is usually convenient to ground the amplifier and use a floating power supply, in spite of its tendency to invite noise. If AC excitation is acceptable, transformer isolation will permit grounding of both power supply and amplifier.

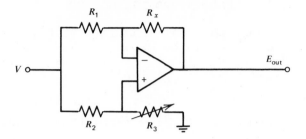

Figure 10-7. An op amp circuit for measuring resistance. This
is sometimes called a "pseudo-bridge."

The requirement of a separate floating power supply can be avoided by substituting an operational amplifier circuit for the Wheatstone bridge, as in Figure 10-7. Circuit analysis shows that the output will
be

$$E_{out} \;=\; V\,\frac{R_1 R_3 - R_x R_2}{R_1(R_3 + R_2)} \qquad (10\text{-}5)$$

If the resistors are adjusted to make E_{out} = 0 (equivalent to balancing a bridge), then the expression in
brackets must be zero, and

$$R_1 R_3 = R_x R_2 \qquad (10\text{-}6)$$

which is identical to the balance expression for the
bridge. This circuit has a distinct advantage when operated in an unbalanced mode, as required for recording changes in resistance. If $R_1 = R_2$, the relation
becomes

$$R_x = R_3 - \frac{E_{out}}{V}(R_1 + R_3) \qquad (10\text{-}7)$$

which predicts a straight line relation between E_{out}
and R_x. For analogous unbalanced operation, the Wheatstone bridge gives a nonlinear response.

TEMPERATURE TRANSDUCERS

Nearly every physical quantity, X, has a temperature variation that can be expressed as a power series:

$$X_T = X_0 [1 + \alpha(T - T_0) + \beta(T - T_0)^2 + \ldots] \quad (10\text{-}8)$$

where X_0 is the value of the variable at temperature T_0
and X_T at temperature T. The temperature coefficient
α can be defined by

$$\alpha = \frac{1}{X_0} \cdot \frac{dX}{dT} \quad (10\text{-}9)$$

Over small temperature excursions, α is nearly constant.
In principle any quantity with a large enough coefficient can be utilized in a temperature transducer.
Those properties of practical importance for this purpose are (1) contact potentials, (2) resistance, and
(3) mechanical dimensions.

The expansion of a liquid, as in the common thermometer, is used more widely than any other method for
temperature measurement, but rarely to give an electronic readout. The change of a linear dimension with
temperature can also be used. A related phenomenon is
the vibration of a plate of crystalline quartz. The
oscillation of this piezoelectric material can easily
be detected electronically, since the frequency is a
function of the temperature. This provides one of the
most stable and sensitive temperature detectors, but
it is quite expensive. Its output is a frequency and
is best measured by digital counting techniques. (Note
that a crystal designed for oscillator service has a
very *low* temperature coefficient.)

The potential developed across the junction between two conductors (the contact potential) is a function of the temperature of the junction, the pair of
conductors constituting a *thermocouple*. For best precision, thermocouples must be connected with one junction at the unknown temperature T_X, the other at a
known temperature T_0. The latter is often 0°C. However, there is no requirement that the reference point
be 0°C, and it may be more convenient to keep it at
room temperature, protected from ambient variations.

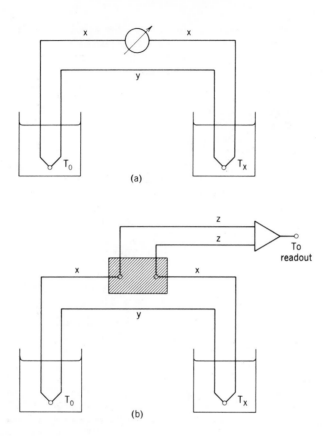

Figure 10-8. A two-junction thermocouple. (a) Simple arrangement, and (b) with an intermediate meter. The shaded area is maintained at a constant temperature. The letters x, y, and z represent different metals.

Figure 10-8 shows two possible connections, using a reference temperature. The designations x and y refer to
two metals, for example copper and the alloy constan-
tan (a widely used couple for the range from -200 to
+350°C). The wires marked z are connections to the
meter (or amplifier) and may be ordinary copper wire,
as long as the two junctions between it and x are held
at the same temperature.

Thermocouples have a wide variety of applications,
from rough monitoring of a furnace to very sensitive

multijunction "thermopiles" for detecting infrared radiation. The thermocouple is inherently a low-voltage,
low-resistance device. Its output can be observed
either as a current or a voltage.
    The change of electrical resistance with tempera-
ture is an important tool in temperature measurement.
Either a metal or a semiconductor can be chosen as the
active component. The metal of choice is platinum because of its long-term stability and chemical inertness.
Nickel has a higher coefficient, but is less reproducible. *Thermistors*, which are semiconductor devices
made by sintering together oxides of various metals,
have a temperature coefficient some ten times larger
than metals (usually of opposite sign). They are less
linear, however, following the relation

$$R_T = R_0 \exp \left(\frac{B}{T} - \frac{B}{T_0} \right) \qquad (10\text{-}10)$$

where B is a constant (of the order of 4000 K^{-1}) and
where T and T_0 must be expressed in Kelvins. Thermistors are excellent for applications involving short-
term measurements and rapid fluctuations in temperature, but are inferior to platinum for work demanding
the highest accuracy.
    Precision resistance thermometry is usually in-
strumented with a "Mueller" bridge, which is especially
designed for this service. This is an expensive re-
search-grade instrument. When rather less precision
is acceptable, a Wheatstone bridge or the circuit of
Figure 10-7 can be used.
    The forward voltage drop across a semiconductor
diode is temperature sensitive and is quite useful as
a sensor. The coefficient is typically -2.5 mV/K for
silicon and somewhat higher for germanium. This temperature effect can be observed and utilized with a *pn*
junction wherever it may appear, in transistors as well
as diodes. In a transistor it may show up as a corresponding change in amplified output. A common application of this effect is in automatic correction for
undesired temperature effects in semiconductor circuits,
since a temperature-produced voltage in one silicon
element is eminently suited to compensate for such a
drop in another. Zener diodes show a temperature dependence of the zener voltage, V_z; the coefficient is

positive for voltages above about 5, negative below,
and very nearly zero just at 5 V. At 10 V the coefficient is typically +6 mV/K. Another temperature-
dependent quantity is the drain current of a FET with
gate and source tied together. This has a temperature
coefficient of about -1%/K. Commercial ICs based on
similar principles are available, for example the AD590
transducer (Analog Devices) gives exactly 1 μA·K^{-1} over
the range from -55°C to +150°C.

OPTICAL TRANSDUCERS

Radiation transfers energy to any material in
which it is absorbed. The energy arrives in the form
of photons with energy E related to the frequency ν
and the wavelength λ by the relation

$$E = h\nu = \frac{hc}{\lambda} \qquad (10\text{-}11)$$

where h is Planck's constant and c is the velocity of
light. If the photons are of short enough wavelength
(great enough energy), they can be observed individually by appropriate detectors as electrical pulses,
and can be counted digitally. If, however, the photons
impinge upon the detector at too great a rate, they
will not be resolved, and the detector will only be
able to give a continuous (analog) output. If the
photons have too little energy (e.g., in the infrared
region), they cannot be counted individually, and the
detector will be able to respond only to the cumula-
tive heat effect. Operating in this mode, detection
of radiation becomes a specialized case of measuring
small elevations in temperature.

PHOTOEMISSIVE DETECTORS

Radiation in the ultraviolet and visible regions
is sufficiently energetic to cause the ejection of
electrons from a metal surface. The number of electrons emitted is given by the number of photons multiplied by a factor (less than 1) which represents the
efficiency of the process. In a practical detector,
the plate (photocathode) is enclosed in a transparent

envelope together with a collector (anode). The anode
is given a positive bias of the order of 100 V to aid
in collecting the electrons, but the number of electrons emitted by the cathode, and hence the current,
is determined by the radiant flux, not by the bias
voltage. (Useful information may possibly be obtained
by operating with a self-generated reverse bias, but
this is not the normal mode.)

There are two modifications of the basic phototube
that provide for amplification of the signal within
the device itself. In one modification the cathode
and anode are immersed in a low-pressure atmosphere of
a noble gas rather than the customary vacuum. Electrons from the cathode collide with gas atoms, and if
the voltage, hence their kinetic energy, is sufficient,
the gas will be ionized. The ions are accelerated by
the bias field, and may produce more ions by subsequent
collisions. The result is that the current is increased perhaps 20 times. Gas phototubes are little used
in critical photometric applications because they tend
to be noisy.

The more significant adaptation of the phototube
is the *photomultiplier* (which is necessarily evacuated).
This device is equipped with a series of added electrodes, called *dynodes*, at successively more positive
potentials. The geometry is such that electrons from
the photocathode impinge on the first dynode with sufficient kinetic energy to cause secondary emission of
three or four electrons per impact. These secondary
electrons are constrained to fall upon the second
dynode, where they produce further emission, and so on.
There is thus a three- or fourfold amplification at
each dynode, and since there may be 10 or 12 dynodes,
the overall amplification may be several million. Because of this internal gain, the photomultiplier is
one of the most sensitive transducers we have. It is
well suited to photon-counting procedures. A fairly
obvious restriction exists against exposure to high
levels of illumination. The conventional circuitry for
a photomultiplier is illustrated in Figure 10-9.

The internal gain of the photomultiplier gives an
added degree of freedom, not present in the simple
vacuum photodiode, that permits operation at constant
anode current. For such service the tube is connected
in the feedback loop of an operational amplifier with
constant input. The amplifier (or its slaved booster)
must be able to supply about 2000 V. Constant current
has advantages from the point of view of power supply

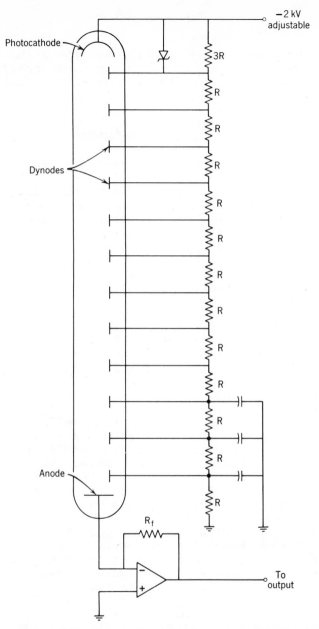

Figure 10-9. Simplified connections for a photomultiplier. The sensitivity can be adjusted by means of the variable voltage supply. The zener diode ensures that the critical voltage difference between cathode and first dynode is maintained at its optimum value. The resistors R are of such a value that the current through the divider chain will be large compared to the tube current, thus avoiding loading effects.

243

safety.   The photomultiplier has been found to be very
effective in this mode, as its response is nearly logarithmic over at least three decades——a valuable feature for spectrophotometry.

SEMICONDUCTOR DETECTORS

The simplest kind of phototransducer in this group
is the *photoconductive cell*.   This is typically made
of CdS, CdSe, or PbS, which can be in the form of a
single crystal, a pressed powder or a vacuum-deposited
film.   The resistance R as a function of illumination
I follows the relation

$$R = kI^m \qquad\qquad (10\text{-}12)$$

where k is a proportionality constant and the exponent
m lies between about -0.4 and -3.0 for CdS prepared by
different methods.   Because of the simplicity of measurement circuitry, the inherent stability, and the
wide dynamic range, photoconductors are very useful in
photographic light meters and simple spectrophotometers.
They are also used in a variety of switching and control applications.
   Another type, also widely used, although less sensitive, contains a *pn* junction.   There are two classes,
known as *photodiodes* and *photovoltaic cells*.   The
chief distinction is that photodiodes are tiny single-
crystal devices, whereas photovoltaic cells are much
larger and polycrystalline.   The latter type, in use
long before the advent of semiconductor electronics in
any other form, includes the self-generating "barrier-
layer" selenium cell.   It has been widely used in col-
orimeters and filter photometers for the visible spectral region.   Its prime advantages are that it requires
no power supply, and that its spectral sensitivity
curve closely approximates that of the eye.   It is less
used in recently designed instruments because of its
rather low sensitivity.   A modern form of the photovoltaic cell is the "solar cell," used in the conversion of solar to electrical energy.
   Photodiodes are limited to very small active areas,
usually less than 1 mm^2, determined by the practical
dimensions of suitable junctions.   The small size reduces drastically the sensitivity, so that photodiodes

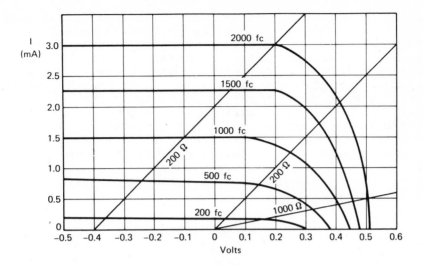

Figure 10-10.   Characteristic curves for a silicon pn-junction photodiode (fc = foot-candles).   (Solar Systems, Inc.)

are used almost exclusively as light-controlled switches in such ON-OFF applications as punched card readers, where small size is an asset.

Figure 10-10 shows the interesting case of a silicon *pn*-junction diode that can be operated with reverse, zero, or forward bias, under short-circuited or open-circuit conditions, or with intermediate load, giving various transfer characteristics. Presumably many other varieties of junction photocells could likewise operate in more than one mode.

The *phototransistor* and *photo-FET* combine the small size and speed of response of the photodiode with the amplifying ability of the transistor or FET to give a device with fairly high gain, but with no specific advantages over a diode-transistor pair.

The energy content of infrared photons is too small to cause electronic transitions in an absorber. It may in some instances be enough to produce charge-separation in a semiconductor, the basis for a photoconductive or photovoltaic device. These types, however, must be cooled with liquid nitrogen (77°K) or even helium (4°K) if the energy of the photon is not to be swamped out by thermal energy of the material of the absorber. The requirement of cooling can be avoid-

ed if the collective rather than individual effects of
photons are utilized. Absorbed radiation causes increased molecular vibration, which is tantamount to an
increase in temperature, and measurement of the radiant flux becomes a measurement of a temperature increase. For high sensitivity, the heat capacity of the
detector must be as small as possible so that the radiant energy available will raise the temperature by a
sufficiently large increment. The radiation is received by a tiny piece of blackened gold foil in contact
with a temperature sensor. The temperature is generally measured with a multiple junction thermocouple,
or with a resistive sensor (for this application, called a *bolometer*).

RADIOACTIVITY TRANSDUCERS

In this section we discuss those detectors that
are useful for radiation with more energy released per
interaction than is available in the optical spectrum.
This includes particulate radiations from nuclear dis-
integrations (α- and β-rays) as well as γ- and x-rays.
These energetic particles and photons can produce ion-
ization in a gas, measureable as a current. A bias
voltage must be applied across a pair of metallic electrodes in a suitable gas. The character of the signal
resulting from a given source will depend on the bias
voltage, as well as the energy content of the incoming
particles or photons, and on their rate of arrival.
If operated at 100 to 200 V, the detector is called an
ionization chamber. The circuit behaves as a current
source controlled by the radiant flux, very much like
the vacuum photodiode; the current is determined solely by the number of ionizing events per second. The
rate at which individual particles can be counted is
limited because of the time required for ions to be
collected by the electrodes. Hence ionization chambers
are used with relatively intense levels of radiation,
where the total current is measured rather than individual pulses.
At higher voltages, the ions are sufficiently accelerated that they acquire enough kinetic energy to
produce additional ions by collision. The ions are
now removed quickly by the impressed field, and the

arrival of each particle is marked by a comparatively
large pulse, proportional in magnitude to the energy of
the primary particle. Furthermore, the response speed
is great enough that more than 10^5 pulses per minute
can be resolved. The bias supply must be well regu-
later, and the amplifier must be strictly linear if
energy proportionality is to be preserved. In this
mode, the device is called a *proportional counter*.

If the voltage is raised still further ($\sim$1000 V,
e.g.), the pulses become larger, and a saturation effect appears, so that the energy content of the ionizing particle no longer controls the pulse size, which
becomes constant. This is the operating region of the
Geiger-Mueller Counter.

While ionization chambers are used mostly for
high-level γ rays, the proportional counter finds its
greatest applications in low-energy counting. The
Geiger counter is easily adaptable to portable operation, and is versatile, being useful for β, γ, and x
rays.

Radioactive particles and photons can also be detected by *scintillation* techniques. The rays are absorbed in a fluorescent liquid or solid, a substance
that gives a tiny flash of light for each particle.
The flashes are observed with a photomultiplier tube
and counted. Photomultipliers give occasional random
pulses not related to an optical input, but merely
statistical fluctuations of electrons in the cathode.
To avoid spurious counts from this source, two photomultiplier tubes may be employed, looking at the same
scintillator. Signals from the two are combined in an
AND-gate, called a *coincidence circuit* (discussed before), so that a count is registered only when both
tubes receive the light from the same flash. This has
long been one of the best systems for counting the activity of relatively weak emitters such as ^{14}C or ^{3}H.

Another transducer for high-energy particles is a
type of photodiode known as a *lithium-drifted germanium*
detector. This has excellent sensitivity and speed,
and might well displace most of the other types were it
not for the inconvenience that it must be kept at or
below liquid nitrogen temperature, even when not being
used. An analogous silicon unit does not require re-
frigeration, but has less sensitivity for penetrating
radiation.

CHEMICAL TRANSDUCERS

Most of the transducers considered in this chapter can be used to measure chemical quantities, but we are presently concerned with those that convert chemical information *directly* into electrical signals. These transducers are more properly called electrochemical, and include types that sense the concentration of ions in solution.

The *concentrations* of ions in solution (more precisely their *activities*) can be measured by means of a pair of specialized electrodes. The solution under test is separated from a standard solution of known activity by a membrane that develops a potential difference between its surfaces depending on the ratio of activities of the specific ion to be measured (Figure 10-11). On each side of the membrane is located a

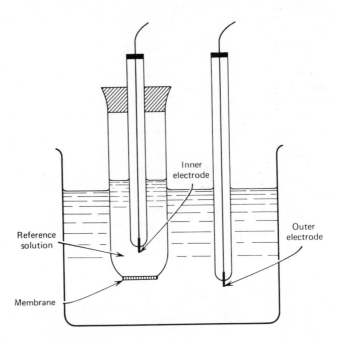

Figure 10-11. Selective-ion electrode cell. The container for the reference solution, together with the membrane and the inner electrode, is commonly referred to as the "working electrode." The outer electrode is called the "reference electrode." Inner and outer electrodes may well be identical.

sensing electrode to measure this potential, E, which
is related to the activity of the ion in the sample
solution, a_M, by the relation

$$E = A + BT \log a_M \qquad (10\text{-}13)$$

where A and B are constants and T is the Kelvin temperature. Membranes are available for many metallic
and a few nonmetallic ions. There are other simpler
electrode systems, useful for related purposes, such
as a platinum wire electrode that senses the oxidation
potential of a solution.

The voltages produced by any one of these transducers seldom exceed 2 V, and may be meaningful when
as low as a few millivolts. Their resistance may run
from a few hundred ohms to many gigohms. A measuring
circuit usable with such high impedance sources is
called a *pH-meter*.

The current-carrying ability of a solution is often very different for DC and AC, and both can give
useful chemical information. The difference arises
mostly because of the energy required to transfer electrons across the boundary from an electrode to ions in
solution (or vice versa): a direct current flowing
through a cell requires that this be accompanied by a
chemical change. Alternating current, on the other
hand, can flow in a circuit including a conducting solution without the actual passage of electrons through
the surface of the electrode. The electrode surface
and a very thin layer of liquid next to it act as a
capacitor, permitting AC to pass, but blocking DC.

The conductance, L, of a solution is the reciprocal of its resistance, as measured with alternating
current (60 to 1000 Hz). Its importance is due to its
direct relation to the total ionic content of the solution. The range of conductances likely to be encountered runs from about 1 to 10^{-7} S (i.e., 1 to 10^7 Ω).
To cover such a wide range, measurement must either be
made on a logarithmic scale, or be covered in a series
of subranges, the latter being the more common. The
instrument is conventionally designed with a Wheatstone
bridge. An op amp circuit has the advantage, however,
that the readout can easily be made linear in conductance units by placing the cell in the input rather
than the feedback.

In contrast to AC conductance is the DC current-
voltage characteristic of a solution. As typical of

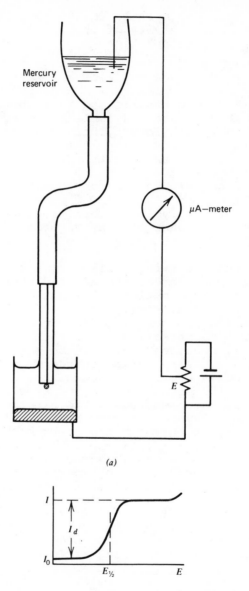

(a)

Figure 10-12. (*a*) Mercury-drop electrode system. (*b*) Current-voltage curve (polarogram) obtained with the cell. The curve shown is a smoothed version of the actual one, which contains serrations due to the periodic falling of the mercury drops.

many electrode systems, consider an inert electrode, such as mercury pool, paired with a very small drop of mercury hanging from the tip of a glass capillary tube (Figure 10-12a). The discipline based on this config-uration is called *polarography*. A variable voltage is impressed across the cell, which contains a solution of electrolytes. As the potential of the drop is made increasingly more negative, current starts to flow. Rather than showing a linear increase with voltage, which would be the case if the solution were replaced by a resistor, the current typically rises in one or more rounded steps (Figure 10-12b). Each step is char-acterized by its height, I_d, which depends on the con-centration of the sensed ion, and by its location, $E_{1/2}$. The $E_{1/2}$ is defined as the potential at which the cur-rent is equal to $I_d/2$; it is characteristic of the species responsible for the step, and can be used in its identification.

In this system the potential need never exceed approximately 2 V, and the current is ordinarily less than 1 mA, conditions suitable for operational ampli-fier control. A circuit is shown in Figure 10-13, in which the two-electrode cell of the previous figure has been replaced with a three-electrode version. Ampli-fier 1 produces a positive-going ramp as biasing volt-age for the cell. Amplifier 2 serves to maintain the reference electrode (marked SCE) at the ramp potential.

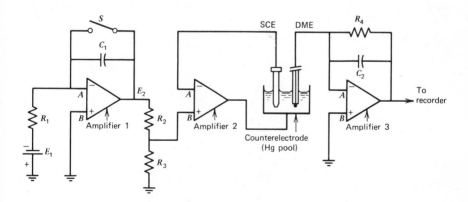

Figure 10-13. An op amp circuit for a three-electrode polarograph.

The third amplifier is a current-to-voltage converter, maintaining the DME electrode at virtual ground while measuring the current through it.

MOVING-COIL METERS

The time-honored D'Arsonval meter movement consists of a lightweight, rectangular coil of wire suspended in the field of a permanent magnet and provided with a pointer. Current is led into and out of the coil through fine spiral springs which also supply the necessary restoring force tending to keep the indicator in its zero position. The moving system may be mounted on pivots or a pair of taut suspension ribbons; the latter is favored for high-sensitivity meters. The movement is basically current sensitive, the torque being proportional to the product of the current and the magnetic field strength.

For AC measurements, a DC meter is sometimes employed, together with a diode rectifier in either half- or full-wave configuration, but the scale must be modified. The "perfect rectifier" circuit of Figure 4-23 or its equivalent must be used if a linear response to low-level AC is needed.

Moving-coil meters for panel mounting are seldom more accurate than ±2% of full scale. Laboratory meters with a long mirrored scale to avoid parallax errors may be accurate to ±1% or better. The relative error of reading increases sharply in the lower third of the meter scale; it is wise to provide range switching at small enough intervals, such as 1, 3, 10, 30, etc., so that the lower part of the scale need not be relied upon when accurate readings are needed.

ACTIVE METERS

This term may be used to denote a combined amplifier and meter, typically provided with switching for multiple ranges and modes of operation. Many commercial analog electronic meters utilize high-gain amplifiers with negative feedback loops containing range-setting resistors. These are in effect operational amplifiers, even though they may not be called by that name. Meters of this type are available in many price levels with correspondingly rated accuracies.

A more recent development is the digital meter.
There are a number of logic systems that convert an analog voltage into its digital equivalent, then display
it numerically. The display can be presented on an
array of *light-emitting diodes* (*LEDs*) that light up at
suitable logic commands to form numerals, as shown in
Figure 9-24.

Digital meters are fast responding, since they
contain no moving parts. Their accuracy can be much
greater than that of comparable analog meters. Some
have a sign indicator (+ or -), and an automatic decimal point. A four-digit instrument can read four significant figures for each included range: for example,
0.1234, 1.234, 12.34, 123.4, or 1234, either sign.

Any type of active meter can be provided with an
electrical output. In the analog case, this permits
connection to a strip-chart recorder. Alternatively,
the meter can be used simply as an amplifier with built-
in range switching. In digital meters the output is
binary in nature, and therefore compatible with digital
computers. The digital voltmeter can become an essential link in data collection and processing systems.

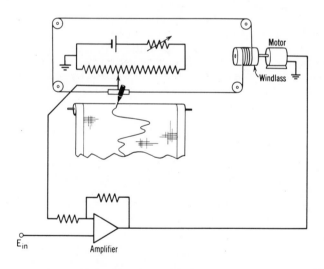

Figure 10-14. Servosystem for a strip-chart potentiometric recorder. The motor-driven carriage moves both the sliding contact
on the voltage divider and the recording pen.

RECORDERS

Recorders, like active meters, can be classified as analog or digital. Most analog recorders produce a graphic display of data in either polar or Cartesian coordinates, by moving a stylus or pen along one coordinate while the recording paper moves along the other. An exception is the x-y recorder, in which the pen moves along both coordinates over a stationary paper. Circular recorders are seldom used in laboratory instruments, but are convenient in many industrial situations.

A servo-actuated self-balancing recorder is shown in Figure 10-14. Loading effects on the system being studied are prevented by the high input impedance. It is insensitive to friction and dust in the moving system, because the energy which produces the motion comes not from the signal but from the feedback.

OSCILLOSCOPES

The cathode-ray oscilloscope is an indispensable tool in the study of any AC or pulse circuitry. Just as in its most familiar embodiment, the home TV receiver, a narrow pencil of electrons is beamed from an electron gun in an evacuated tube to hit a fluorescent screen (Figure 10-15). The trajectory of the electrons in the oscilloscope can be controlled by electrostatic fields applied across two pairs of electrodes called

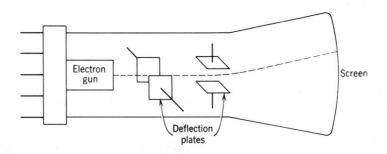

Figure 10-15. A cathode-ray tube for an oscilloscope.

horizontal and vertical deflection plates, and thereby
made to fall at any desired point on the screen. Usually the horizontal deflection plates are energized
with a sawtooth wave, called a sweep.
     The frequency of this sweep is variable, and can
typically be adjusted for times from 1 µsec to 10 sec.
A repetitive signal applied to the vertical deflection
plates, in combination with the sweep, generates an
image of the time variation of the signal. If the
sweep and the signal have the same frequency, the image
retraces itself with each cycle and is held in an apparently stationary display of its waveshape. The
frequency of the signal can be determined from the calibrated dial controlling the sweep generator.
     To be used for nonrepetitive phenomena, an oscil-
loscope must be provided with a trigger circuit that
will allow a single sweep to occur. For example, if
one wishes to examine the shape of the transient produced by the dischange of a capacitor, a sensing circuit should be arranged to trigger the sweep just when
the transient begins, so that the signal and sweep will
be synchronized on the screen. This sort of study is
greatly facilitated by another feature that some scopes
possess—image memory, usually called storage. In this
type, the trace on the screen will remain visible until
erased by a command signal from a pushbutton switch or
its automatic equivalent.
     There are many other varieties of oscilloscopes,
useful in special circumstances. Some types include
a multiplexer, the equivalent of Figure 4-19, and hence
can provide multiple traces so that two or more waveforms can be examined simultaneously. Other scopes are
specifically designed for unusually high frequencies,
for portability, and so on.

PROBLEMS

10-A. Design an operational amplifier circuit to con-
       trol the light intensity of a tungsten lamp by
       means of a photoconductive cell.

10-B. Johnson noise is directly proportional to the
       temperature of a resistor. Design a high-
       temperature thermometer based on this principle
       (a noise thermometer) using op amps. [Refer-

ences: (1) Hogue, NBS Report #3471 (1954), (2)
Anderson and Pipes, *Rev. Sci. Instrum.*, **45**, 42
(1974)]

10-C. Using the circuit of Figure 10-7, where $R_1 = R_2$
= 10 kΩ, and R_3 is a 10-kΩ, 10-turn pot (resolution 0.1%), what would be the minimum tempera-
ture change one could measure with a thermistor
of 10 kΩ (at 25°C) that has a temperature coefficient α = 4.0 %/K?

10-D. Fit Eq. (10-10) to the properties of the thermistor mentioned in Problem 10-C.

                    *   *   *

10-1. Most functions can be expressed as a power series
by means of Taylor's expansion. The expansion
of an exponential function is:

$$e^x = 1 + \frac{x}{1!} + \frac{x^2}{2!} + \frac{x^3}{3!} + \ldots$$

Show over what range the linear approximation of
$f = 1 - \exp(-t/RC)$ is within 10% of the true function.   Hint: let $x = -t/RC$

10-2. Find the transducer equation for a capacitive
gauge where the area of the capacitor varies
with the square-root of the variable.

10-3. Show that the value of S (Eq. 10-4) is approximately 2 if only geometrical effects are considered.

10-4. A single-junction thermocouple with a "thermo-
electric power" of 2 μV/K (Figure 10-16) is directly connected to an ammeter of sensitivity 25
μA full-scale, and resistance R_M = 100 Ω.   The
connections at the meter constitute the cold
junction. The resistance of the couple and its
lead wires totals 10 Ω at room temperature (300°
K) and has a temperature coefficient α = 0.01
K^{-1}. If the thermojunction is raised to 800°K

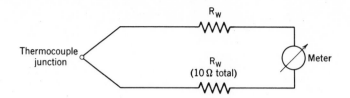

Figure 10-16.   See Problem 10-4.

and the average temperature of the couple and its leads rises to 350°K, what will be the reading of the meter?

10-5. Consider the photomultiplier in Figure 10-9. Assume a conversion efficiency of 0.20 and a gain of 10^7. What is the minimum flux of photons measurable with 5% error, if an op amp with expected drift of 10 nA and 2 mV is used? The charge on an electron is 1.6×10^{-19} C.

10-6. Design an electronic analog voltmeter with ranges in the 1-3-10 sequence from 1 mV to 3-V full-scale. A meter of 50 µA, 1000 Ω, is to be used. The input impedance must be at least 100 MΩ. Use op amp circuitry.

XI

COMBINED ANALOG/DIGITAL CIRCUITS

There are many applications where one cannot be
limited to signals that are only analog or only digital, and a mixture of the two, or a *hybrid*, is essential. Basically we must consider: (a) two separate power supplies must be used, normally ±15 V for the analog and +5 V for the digital part; (b) devices are
needed to convert signals from analog to digital or
vice versa, because of the basically different nature
of the two kinds of signals. This is called *inter-
domain conversion*.

SWITCHES

The simplest device that causes interaction between analog and digital signals is a switch. This device, familiar in everyday life, receives an ON-OFF
command and connects or disconnects analog voltages
accordingly. A few examples of switches are given in
Figure 11-1. Very complex mechanical switches, connecting and disconnecting many lines, are sometimes required, especially in telephone central stations.

Switches can be actuated manually, but of greater
interest are the electrically driven switches called
relays, one example of which is shown in Figure 11-2.
An input signal energizes the coil and attracts the
armature above it, which in turn opens or closes the
switch contacts. A typical relay might require 10 mA

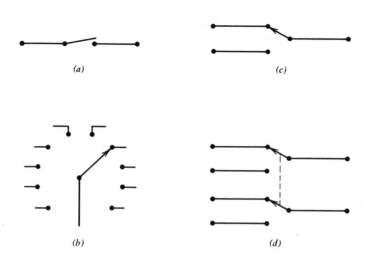

Figure 11-1. Examples of switches: (a) the simplest form, single-pole, single-throw (SPST); (b) a single-pole, 10-throw switch for selection of multiple circuits; (c) a single-pole, double-throw (SPDT); and (d) a double-pole, double-throw (DPDT) switch.

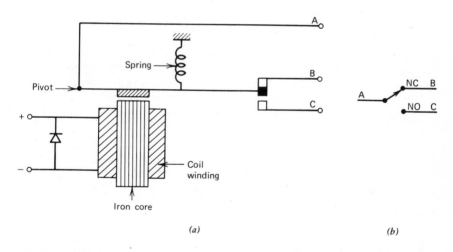

Figure 11-2. (a) A relay, and (b) its switching pattern. The symvols NC and NO stand for "normally closed" and "normally open," referring to the state of the switch contacts when the relay is not excited. Commerical relays vary widely in their voltage and current ratings, AC or DC, and in the number of switch contacts.

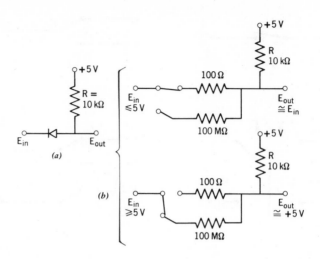

Figure 11-3. (a) A diode switch and (b) its equivalent circuits. By applying the principle of the voltage divider, one can see that the output must be equal either to E_{in} or to +5 V.

at 24 V and execute the switching of several amperes in 20 msec or less. The diode across the winding serves to protect the driving circuit from the high voltages that are generated upon abruptly cutting the current to an inductance (inductive kick).

An interesting type of switch, activated by the signal itself, is the diode. When forward biased, it acts as a closed switch (low resistance) and when reverse biased, as an open switch (high resistance) (Figure 11-3). As long as E_{in} is less than +5 V, the diode will be in its low-resistance state. Current will flow through R and the diode, but essentially all of the voltage drop will appear across R, so that $E_{out} = E_{in}$. When E_{in} becomes greater than +5 V, the diode becomes nonconducting, no current flows through R, and E_{out} = +5 V. The plot in Figure 11-4a demonstrates this relation.

Another form of diode switch with its corresponding plot is shown in Figure 11-4b. Note that the switching action resides in the diode alone; thus in Figure 11-4a E_{out} is either connected directly to E_{in} or to the 5-volt source.

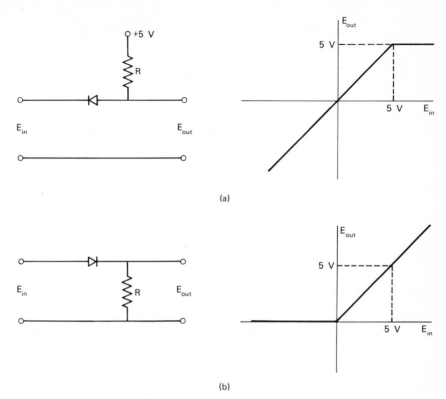

(a)

(b)

Figure 11-4. Two diode switches and their response plots. In either case, zero input gives zero output, and +5 V in gives +5 V out. Circuit (a) acts as a leveler for logic "1," since any higher voltage in gives +5 V out; (b) is a leveler for logic "0," as any negative voltage in produces zero out. In both cases it is assumed that the impedance of the load is very large compared to R, so that the current drawn from the output is negligible.

A variety of such plots can be obtained by utilizing diodes and resistors in various configurations. A number of these are shown in Figure 11-5. In each case the switch is ON when $E_{out} = E_{in}$, that is, when the curve shows a 45° slope, and OFF when E_{out} is not dependent on E_{in}, the plot consisting of a horizontal segment.

It must be emphasized that these plots are idealized. If the switch in its ON state is required to carry signigicant current, the forward resistance of

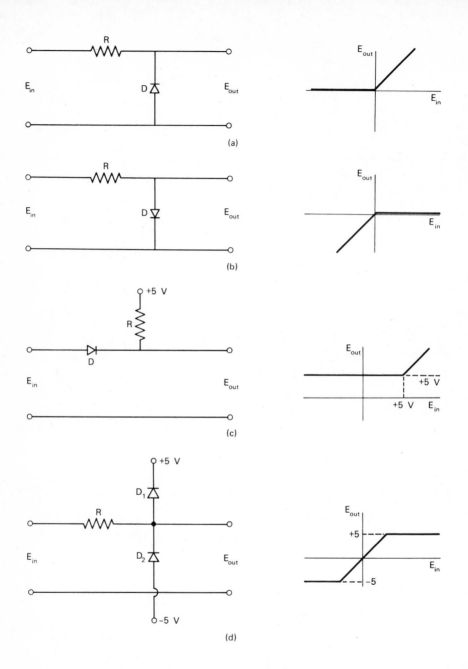

Figure 11-5. Examples of diode switches. The reader is urged to verify the response plots. The circuit in (*d*) is widely used as an overload protection device.

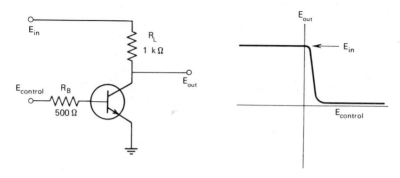

Figure 11-6. A transistor Switch. A silicon switching transistor such as the 2N914 may have an OFF resistance as high as 3 GΩ
and an ON resistance less than 3 Ω, a ratio of about 10^9.

the junction will produce a voltage drop that may not
be negligible. This causes a rounding-off of the angles in the plots. The switching diode 1N3605, for
example, has a forward resistance of 10 Ω at 80 mA,
which will produce a drop of 800 mV. The same diode,
reverse biased, shows a resistance of more than 3 GΩ
(3×10^9 Ω), which may be considered infinite for most
practical purposes.

TRANSISTOR SWITCHES

The transistor, being a three-terminal device, is
more flexible as a switch than the diode. Figure 11-6
shows a basic switching circuit for an *npn* transistor
together with its corresponding plot. As long as the
base is negative (or grounded), the transistor will be
cut off, nonconducting, and because no current can
flow through R_L, there is no voltage drop across it,
and the output, E_{out}, will equal the input voltage, E_{in}.
However, when the base is made positive, the transistor
conducts and the current flowing in the load resistor
causes a voltage drop at the collector. Then E_{out} becomes progressively smaller as $E_{control}$ is increased,
until it reaches a limiting value that is nearly zero.
The sloping portion of the plot contains the "linear" region normally utilized for analog signal ampli-

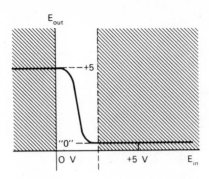

Figure 11-7.   Response of a transistor switch to logic inputs. In the intermediate (unshaded) region, the output is ambiguous.

fication. For switching purposes, this region should be made small by increasing R_L and decreasing R_B as far as possible without interfering with the impedance requirements of the adjacent stages.

The switch of Figure 11-6 can be arranged to accept a digital input at $E_{control}$ and to give digital output, by connecting E_{in} to a +5-V supply. The response can be seen in the diagram of Figure 11-7. Note that the logic levels are not altered in value, although they are interchanged. Moreover, any variation within the shaded area does not affect the output. This is a useful feature in that it permits *restoration* of logic levels. Diode logic switches, on the other hand,

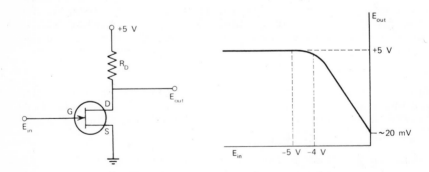

Figure 11-8.   An *n*-channel JFET switch and its transfer plot.

causes shifts of the "0" and "1" levels toward each other, a degeneration of levels that must be corrected by a restoration process.

A FET can be used as a switch in a similar circuit (Figure 11-8). If the gate is made more negative (for an n-channel FET) than about -4 V, the transistor is cut off and E_{out} is equal to +5 V. Grounding the gate causes the transistor to conduct, and the output voltage drops to a few millivolts, the residual potential across the transistor. Therefore, in terms of our logic levels, 0 V in gives 0 V out, but +5 V out requires -5 V in.

PROGRAMMABLE AMPLIFIERS

FET switches can be used very successfully for setting the gain of an operational amplifier. A form of it was shown in Figure 4-19. This is called *gain programming*. It may consist simply in connecting and disconnecting various resistors in either the input or the feedback of an amplifier. An example is shown in Figure 11-9. For each of the four combinations of the digital input lines A and B, there corresponds the connection of one input resistor. Thus "00" gives a gain of 1, "01" a gain of 2, and so on. The usefulness of programmable amplifiers is in automatic operation, and if many stages are to be gain-switched simultaneously,

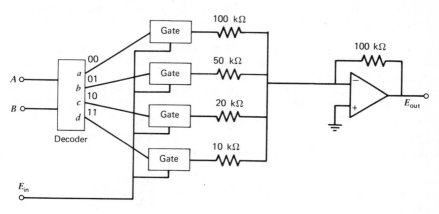

Figure 11-9. A programmable amplifier with gains of 1, 2, 5, and 10.

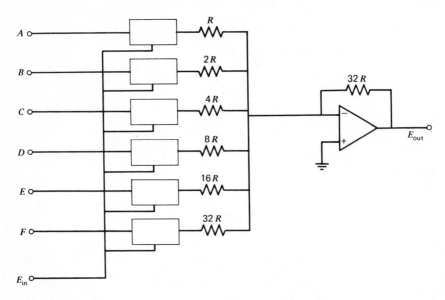

Figure 11-10. A more elaborate programmable amplifier. Logic "1" at any input closes the corresponding switch, thus giving the possibility of 64 different gains from 0 to -63.

such an arrangement can significantly reduce the inconvenience of complex mechanical switching.

The circuit of Figure 11-9 contains a 4-line decoder that can produce logic "1" at only one of the transmission gates at a time, precluding combinations of resistances. If more states are desired, one can use direct logic control. The most useful system employs a set of resistors in a power-of-two sequence $(R, 2R, 4R, 8R, \ldots)$ as shown in Figure 11-10. If we consider the set ABCDEF as a binary number (A being the most significant bit, F the least), then binary 100 000 is equal to decimal 32. This input combination closes the switch in line "A" and gives a gain of $-R_f/R_{in} = -32R/R = -32$. Similarly the binary number 010 000, equal to decimal 16, connects line "B", giving a gain of -16. In other words, the binary number presented to the input becomes equal to the programmed gain of the amplifier. The reader can ascertain that this holds true also if more than one switch is closed. Thus 000 111 gives a gain of -7. The maximum gain is -63, corresponding to binary 111 111, so there are 64, or 2^6 possible gain settings. (Note that binary 000 000 gives a gain of zero.)

DIGITAL-TO-ANALOG CONVERSION

It might appear that proper digital-to-analog conversion would be an impossible task since a given number n of digital lines can only generate 2n values,
whereas analog signals may take an infinity of values
between any two limits. This picture, however, refers
to an ideal representation of analog voltages. For
practical purposes, the question is how many *significant* analog values can be considered to exist in the
given interval. Since analog voltages have an uncer-
tainty Δ caused by statistical deviations, noise, etc.,
two points closer than Δ are not signigicantly different from each other. Consequently an interval of, say,
E volts, has only E/Δ significant points or *channels*,
a situation that can be tackled successfully by a digital approach. To estimate how many digital lines or
bits are needed for error-free operation, one can notice that the number of values of an n-bit binary number is 2^n, whereas the number of channels is E/Δ. Consequently the condition for successfully representing
an analog value by digital means is

$$2^n \; > \; \frac{E}{\Delta} \qquad\qquad (11\text{-}1)$$

This is sometimes described by saying that the number
of combinations (*codes*) must be larger than the number
of channels.

An instrument to perform this operation is called
a digital-to-analog converter, or in short a D/A converter or a DAC. One form of DAC is shown schemati-

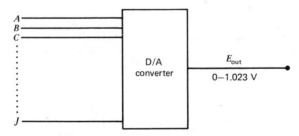

Figure 11-11. A D/A converter. The input has 10 lines, with 2^{10}
= 1024 possible combinations. The full-scale output can usually
be adjusted externally to range the output, for instance, for 0 to
10.00 V.

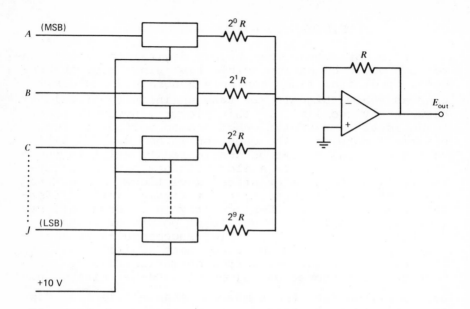

Figure 11-12. An example of a 10-bit DAC. The symbols LSB and MSB refer to the least and most significant bits, respectively. If in place of the +10 V reference, one uses a variable input signal, the device functions as a programmable amplifier when provided with appropriate digital controls.

cally in Figure 11-11. The input is a collection of logical lines that can be thought of as the digital number ABCDEFGHIJ. The numerical value of this number is reflected in the output voltage. There are $2^{10} = 1024$ input codes and they produce corresponding outputs of 0 to 1.123 V in 1 mV steps. In light of the previous discussion, if the output voltage is used in applications where 1-mV resolution is adequate, this can be considered truly analog rather than as a finite (digital) collection of voltages. The internal construction of many D/A converters is comparable to that of a programmable amplifier, as can be seen from Figure 11-12. It is left to the reader to prove that the circuit actually gives 0 to 10 V in 1024 steps, with values represented by the binary number AB...J. It is important to note that there are several other modes of D/A conversion in addition to those discussed above.

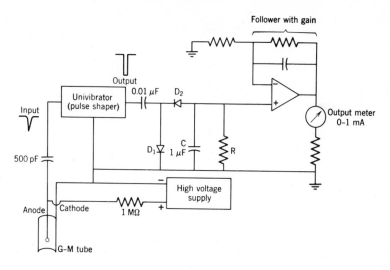

Figure 11-13. A ratemeter, as used with a Geiger counter. (From "techniques of Chemistry," A Weissberger, editor, Vol. I, Part 1B, Wiley-Interscience, N.Y., 1971.)

An example of an instrument that has a built-in D/A converter is the ratemeter used in nuclear and x-ray work (Figure 11-13). Nuclear particles or photons produce individual pulses through a transducer such as a Geiger-Mueller tube. Each pulse contributes an increment of charge to the capacitor marked C, and the resulting voltage is indicated on the meter. The resistor R discharges the capacitor at a rate proportional to the voltage, so that a steady state is attained proportional to the number of counts per unit time. The capacitor and associated components constitute a D/A converter.

ANALOG-TO-DIGITAL CONVERTERS (DIGITIZERS)

There are many ways in which analog signals can be converted to their digital equivalents. One of these is the digital shaft encoder described in Chapter 10.

Another very simple digitizer is a relay used with an analog input and producing a digital output, as

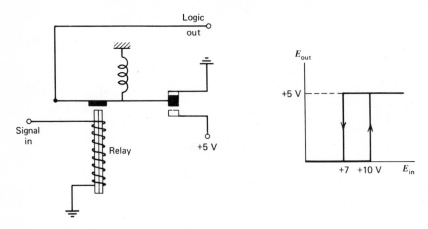

Figure 11-14. A relay configured as a digitizer. In this circuit, the output is logic "1" if the input voltage reaches +10 V, but only shifts back to zero at 7-V input.

shown in Figure 11-14. The response curve of the circuit shows different paths for the 0 → 1 and the 1 → 0 transitions. This effect is called *hysteresis*, and can be very desirable, since the relay latches securely on either value, even in the presence of considerable noise. Without hysteresis, the relay would tend to oscillate ("chatter") between its ON and OFF states.

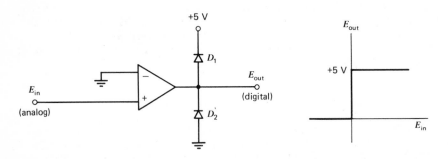

Figure 11-15. An example of a comparator as a digitizer. Diodes D_1 and D_2 limit the output signals to the two extremes, 0 V and +5 V. Diode D_1 can be omitted if D_2 is made a 5-V zener.

Another simple digitizing circuit is based on the voltage comparator introduced in Chapter 4 (Figure 11-15). It answers the query: Is E_{in} positive? By connecting another voltage rather than ground as reference, one can use the circuit to check which of the two voltages is the larger. The circuit shown responds typically within a few microseconds.

In general, digital signals generated from switches, relays or op amps do not have suitable wave forms for direct connection to digital devices. Specifically, the rise time of a conventional integrated circuit logic gate is 10 to 20 nsec, sometimes less,

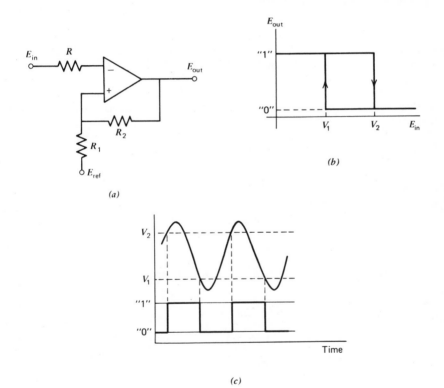

Figure 11-16. (a) The Schmitt trigger, and (b) its wave-shaping function. The trigger can be used to produce a square wave from a sine wave (c). In (a), for minimum offset error, R should equal $R_1 R_2 / (R_1 + R_2)$.

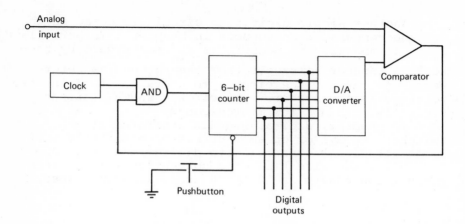

Figure 11-17. A 6-bit analog/digital converter. The pushbutton is pressed to reset the counter, released to start conversion.

and cannot be matched by any mechanical device. This waveform mismatch can be harmful and can be avoided by the use of a special device called a *Schmitt trigger* (Figure 11-16), to shape up the signal. The Schmitt trigger possesses hysteresis, and can accept signals that rise quite slowly.

A more elaborate circuit for A/D conversion is shown in Figure 11-17. A binary counter, driven continuously by clock pulses, is converted to analog by an internal DAC, and the output of this is compared against the input analog voltage. When the comparator changes state, the counter stops, and its output is the digital equivalent of the analog input. This value is held until the reset button is actuated.

PROBLEMS

11-A. Draw the transfer plot for the circuit of Figure 11-18, for $k = 0.5$.

11-B. In Figure 11-19, a gate is compared to an assembly of mechanical switches. Prove that they are equivalent, in the sense that a closed switch corresponds to logic "1" input and an open switch to logic "0." Give the truth table.

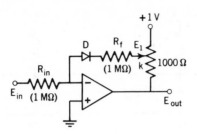

Figure 11-18.   See Problem 11-A.

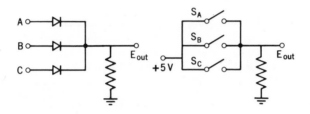

Figure 11-19.   See Problem 11-B.

11-C.   (a)   Design a programmable amplifier, as in Figure 11-9, with gains in a 1-2-5 sequence from 1 to 5000.
(b)   Can you find a method that uses fewer than 12 gates?

11-D.   Design a cascade combination of a programmable-gain amplifier and a 10-bit D/A converter to increase the dynamic range of the analog output beyond the 1023-to-1 afforded by the 10-bit input.   Comment on the precision of this circuit.

11-E.   Consult the literature concerning A/D converters and describe the theory of the dual-slope type.

·*   *   *

11-1.   Construct diagrams corresponding to Figure 11-3 for the switches of Figure 11-20.

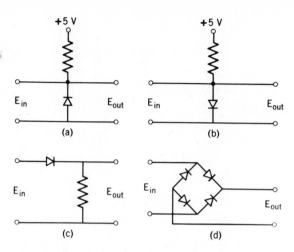

Figure 11-20. See Problem 11-1.

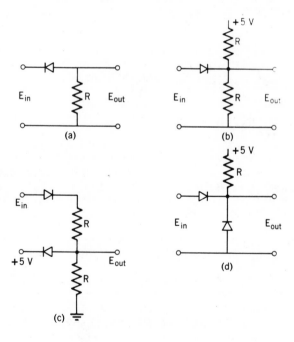

Figure 11-21. See Problem 11-2.

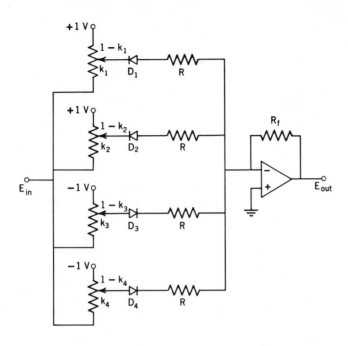

Figure 11-22. See Problem 11-5.

11-2. Sketch the transfer plots for the switches of Figure 11-21.

11-3. Sketch the output waveforms of the circuit of Figure 11-4, if a square wave of ±6 V is applied at the input.

11-4. Sketch the output waveforms for the circuits of Figure 11-5, with a ±6-V square-wave input.

11-5. In Figure 11-22 is shown the circuit for a diode function generator (DFG). (a) Sketch the transfer plot for the circuit if $K_1 = 1/2$, $K_2 = 3/4$, $K_3 = 1/2$, $K_4 = 3/4$, $R = R_f \doteq 1$ MΩ, and the potentiometers are each 1 kΩ. (b) What is the effect of varying the values of the k's? (c) What is the effect of varying the resistors R?

11-6. Design switching circuits using resistors, diodes,

and zeners to implement the transfer plots of Figure
11-23.

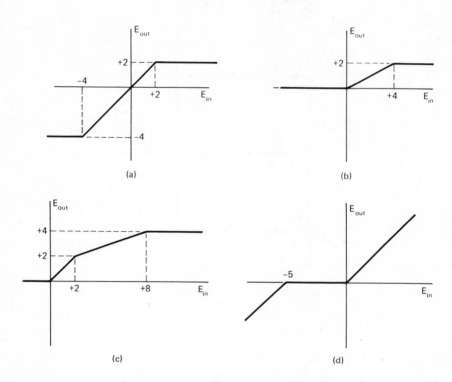

Figure 11-23.   See Problem 11-6.

XII

MATHEMATICAL BACKGROUND

This chapter is intended mostly as a review for the student who has been exposed to the fundamentals of complex algebra and differential equations. The outlines given here are only descriptive introductions to a few mathematical techniques useful in electronics. It is not expected that a student can comprehend the subject matter fully from the very short treatment given. An entire book would be required for each of them. We hope, however, to have abstracted the bare essentials adequately for a student to make some use of them in simple cases, as well as to whet his appetite for a more complete study.

COMPLEX VARIABLES

A complex quantity is defined as a pair of numbers, a and b, usually written as $W = (a + jb)$. The symbol j (or i) is the same as $\sqrt{-1}$. When the properties that $j^2 = -1$, $j^3 = -j$, and so on, are taken into account, most of the operations with conventional (real) numbers are seen to apply to complex numbers as well. In the final result of a complex computation, one normally collects together all terms without j (the real part) and all terms containing j (the imaginary part). For example, one can see that

$$(5 + j3) + (5 - j) = (10 + j2) \qquad (12\text{-}1)$$

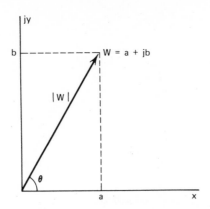

Figure 12-1. Polar representation of a complex number. The symbol W represents the end point of the vector of length $|W|$.

and that

$$(2 + j3) \cdot (2 - j3) = (13 + j0) = 13 \qquad (12\text{-}2)$$

Of special interest in electronics is another description of complex numbers—the *vectorial* or *polar* representation. In Figure 12-1 one can see that the ventor of length $|W|$ which makes an angle θ with the abscissa, defines a point of coordinates a and jb—in other words, a pair representing a complex number.

One can write this equivalence of representation as

$$W = a + jb = |W| \underline{/\theta} = |W| \exp(j\theta) \qquad (12\text{-}3)$$

The last two forms differ only in notation, both containing the same quantities $|W|$ and θ. The relation between various representations is illustrated in Figure 12-2. Both the polar and conventional (cartesian) forms have their advantages. In cartesian form, one can add and subtract with ease, while multiplication and division are particularly simple in polar form. The cartesian addition and subtraction follow the rule

$$(a + jb) + (c + jd) = (a + c) + j(b + d) \qquad (12\text{-}4)$$

with the expected sign changes in subtraction. For multiplication the formula is

$$(a + jb) \cdot (c + jd) = (ac - bd) + j(bc + ad) \qquad (12\text{-}5)$$

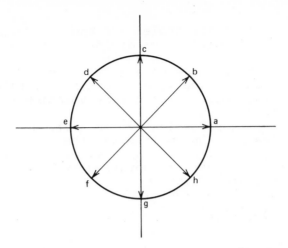

Vector	Cartesian	Polar	Exponential
a	$1 + j0$	$1\ \underline{/0°}$	$1\ \exp\ (j0)$
b	$1/\sqrt{2} + j/\sqrt{2}$	$1\ \underline{/45°}$	$1\ \exp\ (j\pi/4)$
c	$0 + j$	$1\ \underline{/90°}$	$1\ \exp\ (j\pi/2)$
d	$-1/\sqrt{2} + j/\sqrt{2}$	$1\ \underline{/135°}$	$1\ \exp\ (j3\pi/4)$
e	$-1 + j0$	$1\ \underline{/180°}$	$1\ \exp\ (j\pi)$
f	$-1/\sqrt{2} - j/\sqrt{2}$	$1\ \underline{/225°}$	$1\ \exp\ (j5\pi/4)$
g	$0 - j$	$1\ \underline{/270°}$	$1\ \exp\ (j3\pi/2)$
h	$1/\sqrt{2} - j/\sqrt{2}$	$1\ \underline{/315°}$	$1\ \exp\ (j7\pi/4)$

(b)

Figure 12-2. Unit vectors: (a) geometrical relations; (b) various alternative representations. The polar and exponential forms are sometimes written in terms of either degrees or radians, although strictly speaking the exponential form in degrees is improper.

as can easily be verified if the properties of j are taken into consideration. Multiplication in polar form is much simpler:

$$(|W_1|\ \underline{/\theta_1}) \cdot (\ |W_2|\ \underline{/\theta_2}) = |W_1W_2|\ \underline{/\theta_1 + \theta_2} \quad (12\text{-}6)$$

while division obeys the rule

$$\frac{|W_1|\underline{/\theta_1}}{|W_2|\underline{/\theta_2}} = \frac{|W_1|}{|W_2|}\ \underline{/\theta_1 - \theta_2} \quad (12\text{-}7)$$

Observe that the angles add and subtract while the absolute values multiply or divide. For example, multiplication by $2\underline{/60°}$ means an increase in absolute value by a factor of two, with a simultaneous rotation of 60° in the counterclockwise direction. Similarly, division by $1\underline{/90°}$ means a clockwise rotation by 90° with no change in amplitude.

Division in cartesian representation is somewhat more complicated. One multiplies the denominator and numerator by the *complex conjugate* of the denominator. (The complex conjugate is obtained by replacing j by $-j$ wherever encountered.)

$$\frac{a + jb}{c + jd} = \frac{(a + jb)(c - jd)}{(c + jd)(c - jd)} = \frac{(ac + bd) + j(bc - ad)}{c^2 + d^2}$$

$$= \frac{ac + bd}{c^2 + d^2} + j\frac{(bc - ad)}{c^2 + d^2} \qquad (12-8)$$

The interconversion between the two modes of representation can be performed by using the following transformation formulas:

$$x = |W|(\cos \theta) \qquad (12-9)$$

$$y = |W|(\sin \theta) \qquad (12-10)$$

$$x + jy = |W|(\cos \theta + j \sin \theta) \qquad (12-11)$$

$$|W| = \sqrt{x^2 + y^2} \qquad (12-12)$$

$$\theta = \arctan \frac{y}{x} \qquad (12-13)$$

For example, the number $W = 3 - j3$ can be rewritten in polar form by obtaining the absolute value and the angle:

$$\theta = \arctan \frac{-3}{3} = -45° \qquad (12-14)$$

$$|W| = \sqrt{3^2 + 3^2} = 3\sqrt{2} \qquad (12-15)$$

This is a vector of length $3\sqrt{2}$ pointing 45° below the positive x axis. The reverse transformation can be accomplished by calculating

$$x = 3\sqrt{2} \cos (45°) = 3\sqrt{2} \cdot \frac{\sqrt{2}}{2} = 3 \qquad (12\text{-}16)$$

$$y = 3\sqrt{2} \cos (-45°) = 3\sqrt{2} \cdot \frac{-\sqrt{2}}{2} = -3 \qquad (12\text{-}17)$$

By combining the two results, the original number $(3 - j3)$ is obtained.

To understand the usefulness of complex numbers in AC systems, let us consider a circuit at a single frequency. At various points in the circuit voltages will be present, having various amplitudes and phases but all of the same frequency

$$E_1 \sin (\omega t + \phi_1)$$
$$E_2 \sin (\omega t + \phi_2) \qquad (12\text{-}18)$$
$$\dots\dots\dots\dots\dots\dots$$
$$E_n \sin (\omega t + \phi_n)$$

Each of these sinusoidal functions can be described by a rotating vector of angular velocity ω. The entire cluster of vectors (Figure 12-3) rotates with the same velocity and with fixed relative vector orientations as determined by the phase angles. In addition the vectors describing the currents in the circuit will also rotate with the same velocity.

It is advantageous to represent the collection of vectors with their relative orientations as of a certain moment of time, commonly taken as $t = 0$. One can regard it as an instantaneous photograph of the cluster. Such stationary vectors are called *phasors*.

In this representation, the angle between the positive abscissa and each vector is equal to the phase angle, whereas its length represents the corresponding amplitude. It can be shown that all combination rules for phasors are the same as the corresponding rules for complex numbers; the two concepts are nearly identical. Hence the great utility of this representation.

The relationship between the sinusoidal and phasor notations can be written as

$$A \sin (\omega t + \phi) \longrightarrow A \; \underline{/\phi} \qquad (12\text{-}19)$$

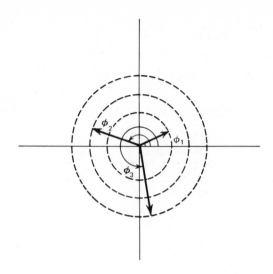

Figure 12-3. A set of rotating vectors at the moment when $t = 0$.

The arrow indicates that the relationship is one of transformation rather than of equality, since the right-hand side does not contain time as a variable. Both voltages and currents can be represented as phasors. It is a widespread convention to express the amplitude as RMS, which simplifies many computations.

IMPEDANCES AND TRANSFER COEFFICIENTS
IN COMPLEX FORM

Since various voltages and currents in phasor form are complex numbers, it follows that their ratios, including impedances and transfer coefficients, will also be complex. This feature, far from complicating the numerical manipulations, makes them more convenient and less prone to mistakes.

It is possible to write directly the impedance of combinations of resistors, capacitors, and inductors by the following rules:

(a)   Impedance of resistors:

$$Z_R = R = R \underline{/0°} \tag{12-20}$$

(b)   Impedance of capacitors:

$$Z_C = \frac{1}{j\,\omega C} = \frac{1}{\omega C}\;\underline{/-90^\circ} \qquad (12\text{-}21)$$

(c)   Impedance of inductors:

$$Z_L = j\,\omega L = \omega L\;\underline{/+90^\circ} \qquad (12\text{-}22)$$

(d)   Series impedances:

$$Z_T = Z_1 + Z_2 + Z_3 +\ldots \qquad (12\text{-}23)$$

(e)   Parallel impedances:

$$\frac{1}{Z_T} = \frac{1}{Z_1} + \frac{1}{Z_2} + \frac{1}{Z_3} + \;\ldots \qquad (12\text{-}24)$$

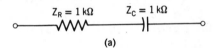

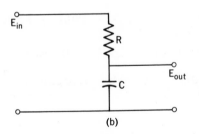

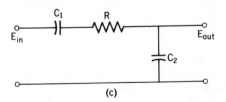

Figure 12-4. *RC* circuits illustrating the phasor method.

The impedance reduces to resistance for nonreactive circuits or for DC signals. It can be considered as an extension of the notion of resistance to a new dimension, where not only amplitudes but phases also are specified. This extension is necessary, since the use of the concept of resistance where impedance should be employed leads to errors.

To illustrate the type of computations that can be made with complex impedances, consider the circuit of Figure 12-4a. The impedance of the series combination is given by

$$Z_T = R + \frac{1}{j\omega C} \qquad (12\text{-}25)$$

Assume that a potential of 100 V is applied and that the frequency is such that the impedance of the capacitor is 1000 Ω. A common error in computing the current would be to assume that the total impedance Z_T is $R +$ $Z_C = (1000 + 1000)$ Ω and compute the current to be 100/2000 = 0.05 A. In reality, if a meter is used to measure the current, a value of 0.071 A would be read.

The serious error in the calculation is due to the use of the concept of resistance instead of impedance. The correct computation is done by obtaining the complex impedance

$$Z_T = R + \frac{1}{j\omega C} = 1000 - j1000 \qquad (12\text{-}26)$$

which can be rewritten in polar form as

$$Z_T = \sqrt{1000^2 + 1000^2} \; \underline{/\text{arctan}\left(\frac{-1000}{1000}\right)}$$

$$= 1414 \; \underline{/-45°} \qquad (12\text{-}27)$$

The correct current can then be computed as

$$I = \frac{E}{Z_T} = \frac{100 \; \underline{/0°}}{1414 \; \underline{/-45°}} = 0.071 \; \underline{/45°} \qquad (12\text{-}28)$$

which is quite different from the tentative result, 0.050 A. An even more unexpected result is seen in the voltages across the resistor and capacitor. "Common sense," or more precisely the unwarranted extension of DC concepts, would indicate that the voltage across the

whole circuit should divide equally between the two impedances as 50 and 50 V. If the voltages are measured,
however, they turn out to be 71 and 71 V. In other
words, two 71-V signals add to a 100-V sum. This unusual result is caused by the fact that the voltages
across the capacitor and resistor are 90° out of phase.
By lengthy trigonometric manipulation it is possible
to show that in fact the two out-of-phase voltages do
add to a 100-V total, but by using phasors the result
comes naturally:

$$E_R = IZ_R = (0.071 \ \underline{/45°}) \cdot (1000 \ \underline{/0°})$$

$$= 71 \ \underline{/45°} = \left(\frac{71}{\sqrt{2}}\right)(1 + j)$$

$$E_C = IZ_C = (0.071 \ \underline{/45°}) \cdot (1000\underline{/-90°})$$

$$= 71 \ \underline{/-45°} = \left(\frac{71}{\sqrt{2}}\right)(1 - j) \qquad (12-29)$$

$$E = E_C + E_R = \left(\frac{71}{\sqrt{2}}\right)(1 + j) +$$

$$\frac{71}{\sqrt{2}}(1-j) = \frac{142}{\sqrt{2}} = 100$$

The phasor notation, although not essential, is far simpler than the conventional description.

The extension of the DC concepts by the use of
phasors is also applicable to transfer coefficients and
gains. The resulting transfer coefficient is complex
and as such contains information about both the amplitude and the phase relations between input and output.
For example, in Figure 12-4b the same pair of impedances shown in Figure 12-4a is now connected as voltage divider. The transfer coefficient E_{out}/E_{in} is giv-
en by the voltage-divider equation

$$\frac{E_{out}}{E_{in}} = \frac{Z_C}{Z_R + Z_C} = \frac{1/j\omega C}{1/j\omega C + R} = \frac{1}{1 + j\omega RC}$$

$$(12-30)$$

From this we can obtain the behavior of the circuit at
various frequencies. Thus at very high frequency the

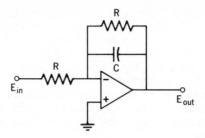

Figure 12-5. An op amp circuit with a complex feedback impedance.

transfer coefficient is zero, and for DC it becomes
unity. The circuit is a low-pass filter.
        Figure 12-4c shows a slightly more complicated
circuit. By the same method, the transfer coefficient
in this case is

$$\frac{E_{out}}{E_{in}} = \frac{1/j\omega C_2}{1/j\omega C_2 + R + 1/j\omega C_1} \tag{12-31}$$

It is left as an exercise for the reader to rewrite
this expression in the conventional format $(a + jb)$.
        Another application of phasors is to determine
the frequency response of operational amplifier circuits. Steady-state sinusoidal voltages are assumed.
The formula to be used is the familiar relation $E_{out}/$
$E_{in} = -Z_f/Z_{in}$. As an example, in the circuit of Figure
12-5 the complex gain (transfer coefficient in complex
form) is

$$\frac{E_{out}}{E_{in}} = - \frac{1}{1/R + j\omega C} \cdot \frac{1}{R} = - \frac{1}{1 + j\omega RC} \tag{12-32}$$

which is identical to Eq. (12-30) for a low-pass filter.

FOURIER SERIES

        Consider a circuit with a frequency-dependent
transfer coefficient $A(\omega)$, as shown in Figure 12-6. If
E_{in} is a mixture of several frequencies, the output is
given simply by the sum of outputs of each sine wave

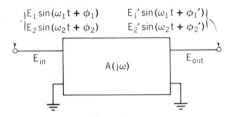

Figure 12-6. Illustration of the principle of superposition. The black box is assumed to contain a linear circuit.

acting individually. This additivity is called the *superposition principle* and extends also to DC voltages, but applies only to linear circuits.

One result of the principle of superposition is that, if a repetitive signal of any form can be expressed as a sum of sinusoidal terms, then the powerful methods of phasor algebra can be applied to it.

In fact, all periodic functions of interest to us* can be expressed as a sum of sinusoidal terms. This is stated by the *Fourier theorem*, which also indicates that the expansion is unique (i.e., for each function there corresponds a single expansion and for each expansion a single function). The theorem applies also to a one-shot signal if it is considered to be a single period of a fictitious repetitive function.

The Fourier expansion can be carried out, for simplicity, in terms of the variable θ defined as $\theta = 2\pi f t$ where t is the time and f the repetition frequency. The expansion is given by the infinite series

$$F(\theta) = \frac{A_0}{2} + A_1 \cos \theta + A_2 \cos 2\theta + \ldots +$$

$$A_n \cos n\theta + \ldots + B_1 \sin \theta + B_2 \sin 2\theta +$$

$$\ldots + B_n \sin n\theta + \ldots \tag{12-33}$$

*Restrictions are that the function should not become infinite nor exhibit an infinity of maxima or discontinuities within one period.

where the A's and B's are numerical coefficients. The components of the expansion consist of a DC term and sines and cosines of θ and its harmonics (multiples). Thus a signal with a repetition rate of 100 Hz, regardless of its form, can always be expressed as a sum of sines and cosines of 100, 200, 300 Hz, and so on, in addition to a DC term.

The process of Fourier expansion is not merely theoretical. By actual measurement, if the 100-Hz signal mentioned above is applied to a tunable filter, the output of the filter is found to contain sine waves of 100 Hz, 200 Hz, and so on, with the amplitudes predicted by the Fourier expansion.

The coefficients can be computed theoretically, if the analytical expression for $F(\theta)$ is known. They are given by the following integrals:

$$A_0 = \frac{1}{\pi} \int_{-\pi}^{\pi} F(\theta)\ d\theta \qquad\qquad (12\text{-}34)$$

$$A_n = \frac{1}{\pi} \int_{-\pi}^{\pi} F(\theta)\ \cos n\theta\ d\theta \qquad\qquad (12\text{-}35)$$

$$B_n = \frac{1}{\pi} \int_{-\pi}^{\pi} F(\theta)\ \sin n\theta\ d\theta \qquad\qquad (12\text{-}36)$$

Thus if $F(\theta)$ is equal to $(\theta^2 + 2\theta)$ between the limits $-\pi$ and $+\pi$, then a fictitious repetitive function can be assumed consisting of repetitions of $F(\theta)$. The expression $(\theta^2 + 2\theta)$ is then substituted in each integral, and the integrals are evaluated. In the process of integration the variable θ disappears when the limits are introduced. A set of numbers A_0, A_1, and so on, is obtained, and one can write

$$(\theta^2 + 2\theta) = \frac{A_0}{2} + A_1 \cos\theta + \ldots + B_1 \sin\theta + \ldots$$
$$(12\text{-}37)$$

There are infinitely many coefficients, a circumstance that may appear to make theoretical computations impossible. As a matter of fact, in many cases the numerical values of the coefficients decrease rapidly with n and can be neglected after a relatively small number of terms. In addition, sometimes a whole group of terms may be zero, such as all sines or all even-numbered cosines.

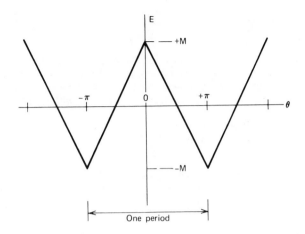

Figure 12-7. A triangular wave. The function $F(t)$ is expressed as a function of $\theta = 2\pi f = 2\pi/T$, where T is the period.

As an example of Fourier expansion, consider the triangular wave shown in Figure 12-7, where the analytical expression for the function is

$$F(\theta) = M - \frac{2\theta}{\pi}\, M \qquad 0 \leq \theta \leq \pi \qquad (12\text{-}38)$$

and

$$F(\theta) = M + \frac{2\theta}{\pi}\, M \qquad -\pi \leq \theta \leq 0 \qquad (12\text{-}39)$$

where M is a constant. The Fourier coefficients can be computed by splitting the integrals into two. Thus

$$A_0 = \frac{1}{\pi} \int_{-\pi}^{0} \left[M + \frac{2\theta}{\pi}\, M \right] d\theta + \frac{1}{\pi} \int_{0}^{\pi} \left[M - \frac{2\theta}{\pi}\, M \right] d\theta$$

$$(12\text{-}40)$$

When performing the integration, the value of A_0 turns out to be zero. This is expected, since there is no DC component.

The A_n terms are obtained by writing

$$A_n = \frac{1}{\pi} \int_{-\pi}^{0} \left[M + \frac{2\theta}{\pi}\, M \right] \cos n\theta \; d\theta \; +$$

$$\frac{1}{\pi} \int_0^\pi \left[M - \frac{2\theta}{\pi} M \right] \cos n\theta \ d\theta \qquad (12\text{-}41)$$

Observe that the two integrals differ only in the sign of θ and in the integration limits. It can be shown that they are indeed equal and it is only necessary to solve one of the two and double it to obtain the answer

$$A_n = \frac{2}{\pi} \int_0^\pi \left[M - \frac{2\theta}{\pi} M \right] \cos n\theta \ d\theta$$

$$= \frac{2M}{\pi} \int_0^\pi \cos n\theta \ d\theta - \frac{4M}{\pi^2} \int_0^\pi \theta \cos n\theta \ d\theta$$

$$= \frac{2M}{\pi n} \sin n\theta \Big|_0^\pi - \frac{4M}{\pi^2 n^2} \int_0^{n\pi} (n\theta) \cos (n\theta) \ d(n\theta)$$

$$(12\text{-}42)$$

The first term has been integrated, and the second rewritten with a new variable, $n\theta$. This requires a change in the upper limit, since when $\theta = \pi$ the new variable $n\theta$ becomes $n\pi$. Substitution of the limits shows the first term to be zero. The second can be estimated by the formula.

$$\int x \cos x \ dx = \cos x + x \sin x \qquad (12\text{-}43)$$

to be found in integral tables. It follows that

$$A_n = - \frac{4M}{\pi^2 n^2} (\cos n\theta - n\theta \sin n\theta) \Big|_0^\pi$$

$$= \frac{4M}{\pi^2 n^2} (1 - \cos n\pi) \qquad (12\text{-}44)$$

The coefficients are zero for $n = 2, 4, 6 \ldots$, since the factor $(1 - \cos n\pi)$ becomes zero. For $n = 1, 3, 5 \ldots$, the quantity $(1 - \cos n\pi)$ equals 2. Consequently the coefficients A_n are

$$A_1 = \frac{8M}{\pi^2} \cdot \frac{1}{1} \qquad A_2 = 0 \qquad A_3 = \frac{8M}{\pi^2} \cdot \frac{1}{9}$$

$$A_4 = 0 \qquad A_5 = \frac{8M}{\pi^2} \cdot \frac{1}{25}, \ \ldots$$

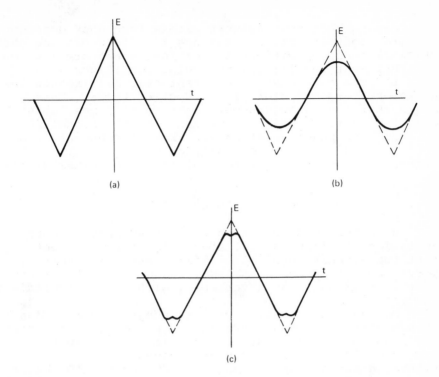

Figure 12-8. (a) Triangular wave; (b, c) approximations obtained by using only one or two terms of the Fourier expansion, respectively.

In addition, it can be shown that all B terms cancel out. It follows that the Fourier series for the triangular wave is given by

$$F(\theta) = A_1 \cos \theta + A_3 \cos 3\theta + A_5 \cos 5\theta + \ldots$$

$$= \frac{8M}{\pi^2} \left[\frac{1}{1} \cos \theta + \frac{1}{9} \cos 3\theta + \frac{1}{25} \cos 5\theta + \ldots \right] \qquad (12\text{-}45)$$

Observe that the terms decrease rapidly. One can obtain a good approximation by retaining only the first 2 or 3 terms (Figure 12-8). Other functions, less similar to a sine wave, may require a larger number of terms.

Coming back to a network with a frequency-dependent transfer coefficient, one can see that to reproduce the original function without distortion, a network (or amplifier) must fulfill the following conditions:

1. The gain must be constant over all frequencies at which there are significant Fourier terms. This means that the bandwidth must be 5 to 10 times larger than the basic repetition frequency of the signal.

2. The phase relation must remain constant. This condition is actually more difficult to fulfill. The effect of phase shift is to distort the original signal.

DIFFERENTIAL EQUATIONS

Both the DC description and the AC phasor representation of circuits have the implicit requirement that the signals involved be time-stable, in other words, that they be in a *steady-state*. In both cases time is eliminated as a variable, with great simplifications.

On the other hand, a circuit subject to a sudden change passes through a transition period, during which its behavior varies, as, for example, when a switch closes or when a signal source is connected to a previously grounded input. Both the DC and the AC phasor approaches are useless in the description of transients, since they cannot cope with time as a variable.

A very general mathematical method for treating transients is by solving the differential equation of the circuit. This somewhat involved method permits the theoretical calculations of steady-state signals as well as transients. We consider here only the case of linear systems consisting of resistors, capacitors, and inductors.

The differential equation of a circuit (more exactly the integrodifferential equation, since it may contain integrals) can be obtained by appropriate combinations of the following relations between voltages and currents:

$$I = C \ \frac{dE}{dt} \qquad\qquad (12\text{-}46)$$

$$E = \frac{1}{C} \int_0^t I \ dt \qquad\qquad (12\text{-}47)$$

$$I = \frac{1}{L} \int_0^t E \, dt \qquad\qquad (12\text{-}48)$$

$$E = L \frac{dI}{dt} \qquad\qquad (12\text{-}49)$$

$$E = IR \qquad\qquad (12\text{-}50)$$

   The first two equations apply to capacitors, the third and fourth to inductors, and the last one is the familiar Ohm's law. For a given circuit one can combine such formulas to obtain the overall differential equation for the system. For example, in the circuit of Figure 12-9 the capacitor is initially charged to E_1 V, but at time zero is suddenly connected to a voltage source E through a resistor R. By inspection of the circuit one can see that, in the steady-state, no current can flow, and the capacitor will be fully charged to E_0 V. In the interim the time dependence of the voltage and current will follow the predictions of the differential equation.

   The equality of voltages around the circuit can be written as

$$E_0 = IR + \frac{1}{C} \int_0^t I \, dt \qquad\qquad (12\text{-}51)$$

To solve this equation we differentiate both sides, obtaining

$$0 = R \frac{dI}{dt} + \frac{I}{C} \qquad\qquad (12\text{-}52)$$

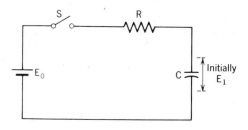

Figure 12-9. An RC circuit illustrating transient and steady-state conditions.

or

$$I = -RC \ \frac{dI}{dt} \tag{12-53}$$

This equation can be shown to have the solution

$$I = A \ \exp \frac{-t}{RC} \tag{12-54}$$

which is easily proved by substitution. The constant A depends on the initial conditions of the circuit. For $t = 0$, A equals I_0, which can be shown to be equal to $(E_0 - E_1)/R$. Consequently the desired expression for the current is

$$I = \frac{E_0 - E_1}{R} \ \exp \frac{-t}{RC} \tag{12-55}$$

From this expression we can obtain the voltage as a function of time by using the relation

$$E = \frac{1}{C} \int_0^t I \ dt + E \tag{12-56}$$

Substituting Eq. (12-55) into Eq. (12-56) gives

$$
\begin{aligned}
E &= \frac{E_0 - E_1}{RC} \int_0^t \exp \frac{-t}{RC} \ dt + E_1 \\
&= - (E_0 - E_1) \ \exp \frac{-t}{RC} \ \Big|_0^t + E_1 \\
&= E_0 \left[1 - \exp \frac{-t}{RC} \right] + E_1 \ \exp \frac{-t}{RC} \tag{12-57}
\end{aligned}
$$

These solutions (12-55 and 12-57) give I and E as functions of time. Thus, as time increases, it appears that the current approaches zero, as expected, while the voltage across C approaches E_0. Similarly, one can show that for $t = RC$ (after one time constant) the current is reduced to $1/e$ of the initial value.

The solution of the differential equation in this example was relatively simple. Circuits with more components pose considerably more mathematical difficulty.

LAPLACE TRANSFORMS

The solution of more complicated differential equations can be greatly simplified by the use of Laplace
transforms. This method, applicable to any linear differential equation, can be considered analogous to the
use of logarithms for multiplication. To clarify the
concept of transformation, let us review the operations
performed in computing the product xy by logarithms.
They are as follows:

1. Transformation of the numbers into their logarithms: $x \longrightarrow \log x$; $y \longrightarrow \log y$.
2. Application of the operational rules: $\log x$
+ $\log y = \log (xy)$.
3. Inverse transformation: $\log(xy) \longrightarrow xy$.

While in the domain of conventional numbers multiplication is lengthy, in the logarithm domain the operation
is replaced by simple addition.

The usefulness of logarithmic transformation becomes even more evident in the process of exponentia-
tion. How, without logarithms, could one evaluate
$3^{4.71}$? In an analogous manner, the rather difficult
manipulation of differentials and integrals can be replaced by operations in the Laplace domain, where only
simple algebra need be performed. The conventional
solutions can be found by inverse transformation. The
overall procedure is as follows:

1. Take the transform of each side of the differential equation.
2. Make necessary algebraic operations.
3. Take the inverse transform.

The Laplace transform is an operation by which
any function of time $F(t)$ is converted into a new function $\bar{F}(s)$. While the variable t is evidently real and
positive, the new variable s is complex. The function
$\bar{F}(s)$, called the transform, has many advantageous properties, and most important being the close similarity
between a function and its derivatives. Such similar-
ities occur only occasionally for time functions.

Laplace transforms obey the following simple relations, in which Λ symbolizes the taking of the transform:

$$\Lambda\ [E(t)] = \overline{E}(s) \qquad \text{or} \qquad \Lambda(E) = \overline{E} \qquad (12\text{-}58)$$

$$\Lambda\left(\frac{dE}{dt}\right) = s\overline{E} - E(0) \qquad (12\text{-}59)$$

$$\Lambda(\smallint\ E\ dt) = \frac{\overline{E}}{s} + \frac{K}{s} \qquad (12\text{-}60)$$

where $E(0)$ is the initial value of E (i.e., at $t = 0$) and K is the value of the integral just after time zero. It should be evident that, apart from initial conditions, differentiation and integration of a function leave the transform unchanged, multiplied by s or $1/s$.

The Laplace transform of a numerical coefficient is the coefficient itself. Thus $\Lambda(kE) = k\overline{E}$. In contrast, the transform of a constant, E_0, representing an initial condition, is E_0/s. This is sometimes called the transform of a step function, since it applies to voltages or currents that did not exist before time zero, when they were switched on.

Let us see how the simple transform relations between a function and its derivatives can be used to advantage in solving a differential equation. Consider again Eq. (12-51)

$$E_0 = IR + \frac{1}{C}\int_0^t I\ dt \qquad (12\text{-}61)$$

Noting that E_0 is switched on at time zero and is therefore a step function, one can write, using Eq. (12-60) with $k = E_1$

$$\frac{E_0}{s} = \overline{I}R + \frac{\overline{I}}{sC} + \frac{E_1}{s} \qquad (12\text{-}62)$$

This equation contains only $\overline{I}$ and the variable s, in addition to various constants. It can be solved for $\overline{I}$ with the functional relation $\overline{I}(s)$ explicitly shown. An inverse transform furnishes the solution of the differential equation (I as a function of time). Thus Eq. (12-62) can be rewritten as

$$\overline{I} = \frac{E_0 - E_1}{s} \cdot \frac{1}{R + 1/sC}$$

which can be further modified to the more convenient form

$$\overline{I} = \frac{E_0 - E_1}{R} \cdot \frac{1}{s + 1/RC} \tag{12-64}$$

In order to find the inverse transform, a table can be consulted, such as that in the *Handbook of Chemistry and Physics* (CRC Press, West Palm Beach, FL). A few examples are given in Table 12-1 (see also Appendix VII), in which the quantity a is assumed to be a constant. From the table one can see that the solution of Eq. (12-64) is

$$I = \frac{E_0 - E_1}{R} \exp \frac{-t}{RC} \tag{12-65}$$

which is identical with Eq. (12-55), obtained by conventional means.

TABLE 12-1

Examples of Laplace Transforms

$F(t)$	$\Lambda F(t) = \overline{F}(s)$
t	$1/s^2$
$\exp(-at)$	$1/(s + a)$
$\frac{1}{a}[1 - \exp(-at)]$	$1/[s(s + a)]$
$t \exp(-at)$	$1/(s + a)^2$
$\frac{1}{a} \sin at$	$1/(s^2 + a^2)$
$\cos at$	$s/(s^2 + a^2)$
Step of amplitude a	a/s

As another example, consider Figure 12-10, where an LC circuit is initially connected to a voltage E_0, so that the capacitor charges to that voltage. The initial conditions must be considered carefully if errors

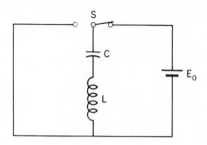

Figure 12-10. An LC circuit in which transients can be induced by means of switch s.

are to be avoided. At time zero, the switch is thrown to the left, shorting the LC combination. The current in the circuit is zero prior to switching and remains zero for an infinitesimal time thereafter. This is so because the inductor opposes any instantaneous increase in current. Consequently, $I(0) = 0$. In contrast, the voltage, initially E_0, suddenly vanishes. This is equivalent to a step of $-E_0$ V. Keeping the initial condition in mind, one can write the differential equation in terms of the sum of the voltages along the circuit

$$E_0 \longrightarrow \quad 0 = \frac{1}{C} \int_0^t I \, dt + L \frac{dI}{dt} \tag{12-66}$$

and its Laplace transform

$$\frac{-E_0}{s} = \frac{\overline{I}}{sC} + Ls\overline{I} \tag{12-67}$$

(Note that it is possible to skip a step and write the Laplace transform directly.) Solving for I gives

$$\overline{I} = - \frac{E_0}{L} \cdot \frac{1}{s^2 + 1/LC} \tag{12-68}$$

The inverse transform is found from Table 12-1, putting $a^2 = 1/LC$

$$I = -E_0 \sqrt{\frac{C}{L}} \cdot \sin \frac{t}{\sqrt{LC}} \tag{12-69}$$

The result indicates no transient, but only a stable sine wave. The negative sign is just an indication of phase.

When more involved differential equations are present, the process of solution requires some manipulations to bring the various polynomials in s to the canonical form found in the tables. If desired, one can invoke the definition of the Laplace transform and evaluate the integrals

$$\overline{F}(s) = \int_0^\infty F(t)\ e^{-st}\ dt \qquad\qquad (12\text{-}70)$$

$$F(t) = \frac{1}{j\,2\pi}\ \int_{a-j\infty}^{a+j\infty} [e^{-st}\ \overline{F}(s)]\ ds \qquad (12\text{-}71)$$

The expression of the inverse transform is somewhat forbidding, even though it may not be as difficult as it appears. It consists of evaluating the line integral in the complex plane along the "vertical" parallel to the j axis at a distance a, from $-\infty$ to $+\infty$.

For illustration, let us evaluate the direct transform of $F = \exp(-at)$:

$$F = \int_0^\infty \exp(-at)\ \exp(-st)\ dt$$

$$= \int_0^\infty \exp[-(s+a)t]\ dt$$

$$= -\ \frac{1}{a+s}\ \exp[-(s+a)t]\Big|_0^\infty = \frac{1}{a+s}$$

$$(12\text{-}72)$$

which is the transform indicated in the table.

s-DOMAIN IMPEDANCE

In the process of generalizing the concept of resistance, we have defined the notion of complex impedance, which represents the ratio of voltage to current for reactive circuits in the presence of steady-state DC or AC. This extension to AC is completely satisfactory, especially in light of the Fourier theorem.

There are applications, however, where this approach cannot be used, namely in the description of transients. Such transients exist in reactive systems

as the result of a sudden change, the most common being
those occuring when a signal source is connected to the
system. For instance, after applying a sine wave to
the input of a circuit, the output signal may be small
in amplitude at the beginning and increase progressively to its steady state. After a sufficiently long
time has elapsed, the value predicted by the phasor
method will be attained.

A very general concept, able to cope with steady
state as well as with transients, is the *s-domain impedance*. This approach is derived from the differential equation for the particular circuit, and as such
it does not have the limitations of the phasor treatment.

For a combination of resistors, inductors, and
capacitors one can write a differential equation in
terms of the voltages across each component. For example, if a series RCL circuit is considered, the equation for the overall voltage E is

$$E = IR + L \frac{dI}{dt} + \frac{1}{C} \int_0^t I \, dt \qquad (12\text{-}73)$$

The Laplace transform of this equation could have been
written directly, denoting each resistor by R, each inductor by sL, and each capacitor by $1/sC$, as follows:

$$\overline{E} = \left[R + Ls + \frac{1}{sC} \right] \overline{I} \qquad (12\text{-}74)$$

The quantity in parentheses is called the s-domain
impedance, or simply the impedance. It is identical in
form to the complex impedance $(R + j\omega L + 1/j\omega C)$, but
with a somewhat different meaning. This identity of
form ($j\omega$ replaced by s) makes the s-domain impedance
an extension of the conventional complex impedance.

Both types of impedance are complex quantities,
but the phasor describes relations between vectors,
whereas the s-impedance describes differential equations. In addition, the latter is more widely applicable than the former in that it covers transients. In
order to express the s-impedance one need not write the
differential equation explicitly, but can set it up by
the use of the following simple rules:

    1. The impedances of single elements are, respectively, R, sL, and $1/sC$.

    2. Series and parallel combinations are given by the well-known rules for combinations of impedances.

As an example, the RC series circuit will have an impedance

$$\bar{Z} = R + \frac{1}{sC} = \frac{RCs + 1}{sC} \qquad (12\text{-}75)$$

If a signal is applied, the response can be obtained by using Ohm's law:

$$\bar{I} = \frac{\bar{E}}{\bar{Z}} \qquad (12\text{-}76)$$

Assuming the signal to be a step of amplitude E_0, one can write

$$\bar{I} = \frac{E_0}{s} \cdot \frac{sC}{1 + RCs} \qquad (12\text{-}77)$$

(This particular transform was encountered earlier, in a case where initial conditions were present.) Setting $E_1 = 0$, we can use the solution of Eq. (12-66):

$$I = \frac{E_0}{R} \exp \frac{-t}{RC} \qquad (12\text{-}78)$$

This equation gives a complete description of the transient. In addition, the steady state can be obtained if t is taken to be infinite. In this case, $I \neq 0$, a result expected from the classical notion of impedance.

To illustrate the application of Laplace transforms to AC systems, let us consider the case of a single capacitor. Consulting Table 12-1, we see that the sinusoid $(1/\omega) \sin \omega t$ has the transform $1/(s^2 + \omega^2)$. The signal applied, $E_0 \sin \omega t$, can be written as $\omega E_0 \cdot [(1/\omega) \sin \omega t]$, with the transform $\omega E_0/(s^2 + \omega^2)$. The current transform is thus given by

$$\bar{I} = \frac{\bar{E}}{\bar{Z}} = \frac{\omega E_0/(s^2 + \omega^2)}{1/Cs} \qquad (12\text{-}79)$$

or

$$\overline{I} = E_0 \omega C \ \frac{s}{s^2 + \omega^2} \qquad (12\text{-}80)$$

The inverse transform, from Table 12-1 is

$$I = \omega C E_0 \cos \omega t \qquad (12\text{-}81)$$

which indicates that the current leads the voltage by
90° and has an amplitude of $\omega C E_0$. The same result is
obtained from phasor computations:

$$I = \frac{E_0}{1/j\omega C} = \omega C E_0 \underline{/90°} \qquad (12\text{-}82)$$

It is interesting to note that there is no transient
present in this system. The assurance that this is the
case could not have been obtained from phasor techniques.

TRANSFER FUNCTIONS

The transform method is applicable not only to
impedances, but also to any other ratios of voltages
and currents, such as gains, and transimpedances. This
represents an extension in the use of the concept of
transfer coefficient to a new quantity in the s-domain,
called the *transfer function*, defined as the ratio of
output to input transforms.

An example of the use of the transfer function is
the s-analog of the voltage divider equation. For the
circuit of Figure 12-11 the attenuation is described
by the transfer function G:

$$G = \frac{E_{out}}{E_{in}} = \frac{\overline{Z}_L}{\overline{Z}_T} = \frac{sL}{R + sL} \qquad (12\text{-}83)$$

where $\overline{Z}_T$ is the total s-impedance.

Such expressions are very useful for describing
transient processes. For steady-state AC, the phasor
form is usually more convenient; as was pointed out, it

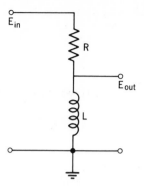

Figure 12-11. An *RL* voltage divider, assumed to be connected to a high-impedance load.

has the same form except for the replacement of s by $j\omega$.

Consider now the effect of a step E_0/s applied to the circuit of Figure 12-11. The transform of the output is

$$\overline{E}_{out} = \frac{E_0}{s} \cdot \frac{sL}{R + sL} = E_0 \frac{1}{(R/L) + s} \qquad (12\text{-}84)$$

By consulting the table of inverse transforms and setting $a = R/L$, one obtains

$$E_{out} = E_0 \exp \frac{-RT}{L} \qquad (12\text{-}85)$$

The output is initially equal to the input, but decreases exponentially to its steady-state value of zero. The transfer function is **particularly** useful in connection with operational amplifier circuits. Consider an amplifier in a single-ended configuration with one feedback and one input impedance. The impedances themselves can be made up of any number of resistors and capacitors. The input current transform is

$$\overline{I}_{in} = \frac{\overline{E}_{in}}{\overline{Z}_{in}} \qquad (12\text{-}86)$$

whereas the feedback current is

$$\overline{I}_f = \frac{\overline{E}_{out}}{\overline{Z}_f} \qquad (12\text{-}87)$$

Writing $\overline{I}_{in} = -\overline{I}_f$, it follows that

$$\frac{\overline{E}_{in}}{\overline{Z}_{in}} = \frac{-\overline{E}_{out}}{\overline{Z}_f} \qquad (12\text{-}88)$$

or

$$\overline{E}_{out} = -\frac{\overline{Z}_f}{\overline{Z}_{in}} \overline{E}_{in} \qquad (12\text{-}89)$$

This is the transform equivalent of Eq. (4-5). It en-
ables us to compute the response of the circuit to any
step input (this would normally be referred to as the
response to a DC signal). Thus the circuit of Figure
12-12a has a transfer function

$$G = -\frac{R + (1/sC)}{R} \qquad (12\text{-}90)$$

Such an operational amplifier circuit, when connected
to a DC source of E_0/s V gives an output of

$$\overline{E}_{out} = G\overline{E}_{in} = -\frac{E_0}{s} \cdot \frac{R + 1/sC}{R}$$

$$= -\frac{E_0}{s} - \frac{E_0}{RCs^2} \qquad (12\text{-}91)$$

By inverse transform it follows that the output is
given by a step $(-E_0)$ at time zero, in addition to a
ramp $(E_0 t/RC)$ generated by the second term.

The transfer functions for the rest of the circuits in Figure 12-12 can easily be shown to be:

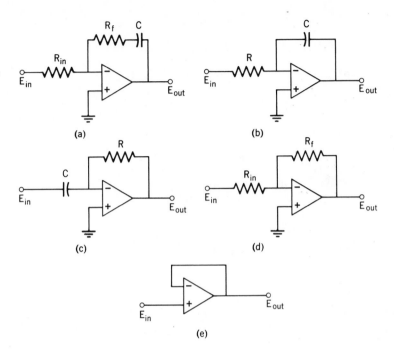

Figure 12-12. Typical op amp circuits illustrating the transfer function.

(b)   The integrator $G = -1/RCs$ (12-92)
(c)   The differentiator $G = -RCs$ (12-93)
(d)   The inverter $G = -R_f/R_{in}$ (12-94)
(e)   The follower $G = 1$ (12-95)

The transient response can be obtained directly by inverse transformation. For example, an integrator connected to an input E_0 gives as output

$$\bar{E}_{out} = \left\{ \frac{E_0}{s} \right\} \left\{ \frac{1}{RCs} \right\} = \left\{ \frac{E_0}{RC} \right\} \left\{ \frac{1}{s^2} \right\}$$ (12-96)

which has as its inverse transform

$$E_{out} = \left(\frac{E_0}{RC} \right) t$$ (12-97)

PROBLEMS

12-1. Perform the following operations:

(a) $(1 + j2) + (6 - j7) \cdot (2 + j2)$

(b) $\left[\dfrac{1 + j}{1 - j} \right] \cdot (2 - j2)$

(c) $(3 + j3)^2$

(d) $\left[\dfrac{1 + j2}{j} \right]^2$

12-2. Carry out the following operations:

(a) $(4 \ \underline{/67°}) \cdot (2 \ \underline{/-30°})$

(b) $(4 \ \underline{/67°}) / (2 \ \underline{/-30°})$

(c) $(31 \ \underline{/90°}) - (6 \ \underline{/45°})$

(d) $(A \ \underline{/\theta})^{2 \cdot 5}$

12-3. Convert the following to their equivalents in polar or cartesian coordinates, as appropriate:

(a) $(-6 - j6)$

(b) $(4 - j4\sqrt{3})$

(c) $(11 \ \underline{/60°})$

(d) $(11 \ \underline{/-90°})$

12-4. Express the impedances of the circuits of Figure 12-13 in complex form.

12-5. Consider a 12-W light bulb of 12 V rating, connected in series with a capacitor. (a) What value of capacitor must be used to permit direct connection to the 115-V 60-Hz power line? (b) What is the minimum voltage rating of the capacitor? (c) How much power, defined as the real part of the EI product, is dissipated in the capacitor?

12-6. Calculate the transfer coefficients for the circuits of Figure 12-14.

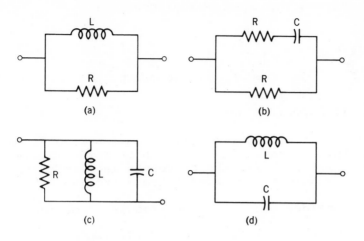

Figure 12-13.   See Problem 12-4.

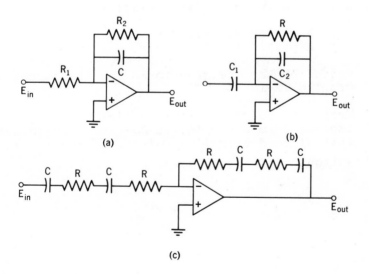

Figure 12-14.   See Problem 12-6.

12-7. Expand the square-wave function:

$$E = E_0 \qquad 0 < E < \pi$$

$$E = -E_0 \qquad -\pi < E < 0$$

in a Fourier series.

12-8. Show that, if a function has a period T, and a Fourier expansion is taken over the interval $-T/2$ to $+T/2$, the corresponding limits for the variable θ (Eq. 12-34) are $-\pi$ to $+\pi$.

12-9. Consider a signal expressed as a Fourier series

$$E = 3 + (1/3)\sin \omega_1 t + (1/9)\sin 2\omega_1 t$$

What will be the output signal if E is applied to a network whose transfer coefficient is $1/j\omega C$?

12-10. Write the integro-differential equations for:

(a) A series RCL circuit.
(b) A parallel RCL circuit.

12-11. Write the Laplace transforms for the circuits of Problem 12-10.

12-12. Write and solve the differential equation in the Laplace form for a series RL-circuit with zero initial conditions; a voltage source E_1 is switched in at time zero.

ANSWERS TO NUMBERED PROBLEMS

2-1. If $F(\theta)$ is $\sin \theta$, we can write

$$E = E_0 \sin \theta$$

and since we wish to average the square, we write further

$$E^2 = E_0^2 \sin^2 \theta$$

Then

$$\overline{E^2} = \frac{1}{2\pi} \int_0^{2\pi} E_0^2 \sin^2 \theta \, d\theta$$

$$= -\frac{E_0^2}{2\pi} \left[\frac{1}{4} \sin 2\theta - \frac{1}{2} \theta \right]_0^{2\pi}$$

$$= \frac{E_0^2}{2\pi} \left(\frac{2\pi}{2} \right) = \frac{E_0^2}{2}$$

Hence

$$\sqrt{\overline{E^2}} = \frac{E_0}{\sqrt{2}}$$

2-2.
$$\overline{E^2} = \frac{1}{2\pi} \int_0^\pi (+V)^2 \, d\phi + \frac{1}{2\pi} \int_\pi^{2\pi} (-V)^2 \, d\phi$$

$$= \frac{1}{2\pi} 2\pi V^2 = V^2$$

or

$$E_{RMS} = V$$

2-3. (a) $2\pi f = \omega = 240$ rad/sec

$$f = \frac{240}{2\pi} = 38.2 \text{ Hz}$$

(b) $\phi = \phi_1 - \phi_2 = 180° = \pi$ rad (the current leads)

2-4. $I_{pp} = 2(0.1) = 0.2$ A

$$I_{RMS} = (0.71)(0.1) = 0.071 \text{ A}$$

$$E_{pp} = 2(10) = 20 \text{ V}$$

$$E_{RMS} = (0.71)(10) = 7.1 \text{ V}$$

2-5. $P = E_{RMS} I_{RMS} \cos (\phi_1 - \phi_2)$

$$= \frac{0.3}{\sqrt{2}} \cdot \frac{3.0}{\sqrt{2}} \cos \frac{\pi}{2} = 0$$

The power factor is zero.

2-6. (a) 20 log 0.01   = -40 dB
      (b) 20 log 0.1   = -20 dB
      (c) 20 log 1   = 0 dB
      (d) 20 log 100   = +40 dB
      (e) 20 log 2   = +6 dB
      (f) 20 log 3.142 = 20(0.497) = 9.94 dB
      (g) 20 log 90   = 20(1.95) = 39.0 dB

2-7. $20 \log \dfrac{E_2}{E_1} = -20;$ $\dfrac{E_2}{E_1} = 0.100$

   $20 \log \dfrac{E_3}{E_2} = -3;$ $\dfrac{E_3}{E_2} = 0.708$

   $\dfrac{E_3}{E_1} = \left(\dfrac{E_3}{E_2}\right)\left(\dfrac{E_2}{E_1}\right) = (0.708)(0.1) = 0.0708$

   $20 \log 0.0708 = 20\,(\overline{2}.85) = -23$ dB   <u>Q.E.D.</u>

2-8.   60 dB corresponds to the ratio 1000:1
       10 dB corresponds to the ratio 3.16:1
       <u>3 dB</u> corresponds to the ratio 1.41:1
       73 dB corresponds to (1000)(3.16)(1.41) = 4456
                                         = 73.0 dB   <u>Q.E.D.</u>

2-9. $R = 100\ \Omega,\ 2$ W

   $P = EI = \dfrac{E^2}{R}$

   $E_{max} = \sqrt{RP} = \sqrt{200} = 10\sqrt{2} = 14.1$ V

2-10. $\text{NEP} = \dfrac{\text{noise output}}{\text{gain}} = \dfrac{e_n^2}{R_{out} \cdot A_V}$

   $= \dfrac{(10^{-3})^2}{(1000)(100)} = 10^{-11}$ W

2-11. $\overline{E_n^2} = B^2 \Delta f;$ $B^2 = 0.1$ mV·Hz$^{-1/2}$ $= 10^{-4}$ V·Hz$^{-1/2}$

   E_n (RMS) $= B\sqrt{\Delta f}$ $E_s = 0.1$ V

   (a) $\Delta f = 100$ Hz

   $S/N = \dfrac{0.1}{10^{-4} \cdot \sqrt{100}} = 100$

   (b) $\Delta f = 10$ Hz

   $S/N = \dfrac{0.1}{10^{-4} \cdot \sqrt{10}} \cong 320$

(c) $\Delta f = 1$ Hz

$$S/N = \frac{0.1}{10^{-4} \cdot \sqrt{1}} = 1000$$

3-1.

$$\frac{1}{Z_t} = \frac{1}{100} + \frac{1}{151}$$

$$Z_t = \frac{(151)(100)}{151 + 100} \cong 60 \ \Omega$$

3-2. $E_{in} = 10$ V $E_{out} = 10$ mV $= 0.010$ V

$$\frac{E_{out}}{E_{in}} = \frac{0.010}{10} = 0.0010$$

$$R_{in} = \frac{E_{in}}{I_{in}} > \frac{10}{0.0010} = 10 \ k\Omega$$

$$R_{out} < 20 \ \Omega$$

Choose, for example, $R_{in} = 15$ kΩ, $R_{out} = 15$ Ω.
Then $R_{in} = R_1 + R_2 = 15$ kΩ and $R_{out} = R_2 = 15$ Ω.
Whence $R_1 = 15000 - 15 = 15$ kΩ, approximately, and
$R_2 = 15$ Ω.

3-3. $I = \dfrac{E}{Z} = \dfrac{E}{R} = \dfrac{10}{100} = 0.1$ A

3-4.

$$E_{Th} = E_{in} \left[\frac{R_2}{R_1 + R_2} \right]$$

$$I_{sc} = E_{in} \left(\frac{1}{R_1} \right) \quad (I_{sc} \text{ means short-circuit current.})$$

$$Z_{Th} = \frac{E_{Th}}{I_{sc}} = \frac{R_2}{R_1 + R_2} R_1 = \frac{R_1 R_2}{R_1 + R_2}$$

Note that this is the parallel combination of
R_1 and R_2.

3-5. The equation is $\omega = 1/RC$. Choose any convenient value of C, such as 0.001 μF, then, $R = 1/\omega C = 1/(10^6)(0.001 \times 10^{-6}) = 1000$ Ω.

3-6. Assume the frequency of E_{in} to be ω rad/sec.

$$E_{Th} = E_{in}\left(\frac{Z_C}{Z_C + R}\right) = E_{in}\left(\frac{1/\omega C}{1/\omega C + R}\right) = E_{in}\left(\frac{1}{1 + \omega RC}\right)$$

$$I_{sc} = E_{in}/R$$

$$Z_{Th} = E_{Th}/I_{sc} = \frac{E_{in}}{1 + \omega RC} \cdot \frac{R}{E_{in}} = \frac{R}{1 + \omega RC}$$

3-7. In open-circuit operation, $E_{out} = E_{Th} = I_n R_n$. In short-circuit conditions, $I_{sc} = E_{Th}/R_{Th} = I_n$. Hence the Thevenin equivalents are

$$R_{Th} = \frac{E_{Th}}{I_n} = R_n$$

$$E_{Th} = I_n R_n$$

(It can be shown that such a current source with shunt resistor exists as equivalent to every Thevenin circuit, hence equivalent to every circuit containing only power sources and impedances; this is the so-called *Norton theorem*.)

3-8. (a) The input resistance is given by the series combination of R_1 with the parallel system of R_2 with R_3 and R_L

$$R_{in} = \frac{660 \times 2970}{660 + 2970} + 60 = 540 + 60 = 600 \ \Omega$$

The input current splits between R_2 and ($R_3 + R_L$) in inverse ratio to the resistances

$$\frac{I_L}{I_2} = \frac{R_2}{R_3 + R_L} = \frac{2970}{660} = 4.50$$

But since $I_2 = I_{in} - I_L$, we can write

$$I_L = 4.50\, I_2 = 4.50\, (I_{in} - I_L)$$

from which

$$\frac{I_L}{I_{in}} = \frac{4.50}{5.50} = 0.818$$

The decibel ratio is 20 log 0.818 = -1.7 dB.

The voltage attenuation is given by E_{out}/E_{in}, where

$$E_{out} = I_L R_L \quad \text{and} \quad E_{in} = I_{in} R_{in}$$

so that

$$\frac{E_{out}}{E_{in}} = \frac{I_L R_L}{I_{in} R_{in}} = \left(\frac{4.50}{5.50}\right)\left(\frac{600}{600}\right) = 0.818$$

Again the decibel ratio is -1.7 dB. The power attenuation is given by P_L/P_{in}:

$$\frac{P_L}{P_{in}} = \frac{I_L^2 R_L}{I_{in}^2 R_{in}} = (0.818)^2 \left(\frac{600}{600}\right) = (0.818)^2$$
$$= 0.669$$

The decibel ratio is 10 log $(0.818)^2$ = -1.7 dB. (Note that the equality of the three decibel ratios is not general, but results from the equality of input and output impedances.)

(b) Suppose that R_L in Figure 3-42 is replaced by another similar three-resistor network and a terminating load. Since the new unit has a 600 Ω impedance, it can replace the former 600 Ω load without change. An obser-

ver at the input cannot distinguish by electrical measurements whether the load is connected directly or through any number of
similar attenuators. An observer at the
load sees voltage, current, and power attenuation of 1.7 dB per unit. Analogous networks with input impedance independent of
length, especially those designed for high-
frequency signals, are known as *transmission
lines*.

3-9.   The slope of 40 dB/decade means that the ratio of
voltages is the square of the ratio of frequencies:

$$\frac{E_{out}}{E_{ref}} = \left(\frac{f_0}{f_2}\right)^2$$

The second harmonic (at 4000 Hz) thus corresponds
to

$$\frac{E_{out}}{E_{ref}} = \left(\frac{3000}{4000}\right)^2 = 0.56$$

The second harmonic content is given by

(0.56)(2/10) = 0.112 or 11.2%

For the third harmonic (6000 Hz) the initial content is 10%. This is altered by the factor

$$\left(\frac{3000}{6000}\right)^2 = \frac{1}{4}$$

giving

(0.25)(10%) = 2.5%

3-10. The initial S/N = 100/10 = 10.

$$f_0 = \frac{1}{2\pi RC} = \frac{1}{(6.28)(10,000)(0.0001)} = 0.159 Hz$$

The attenuation for a simple RC filter is 6 dB/
octave, so that we can write

$$\frac{f}{f_0} = \frac{E_{in}}{E_{out}}$$

$$\frac{60}{0.159} = \frac{E_{in}}{E_{out}} = 377$$

$$E_{n(out)} = \frac{10}{377} = 0.0265 \text{ mV}$$

Hence the final S/N is

$$S/N = \frac{100}{0.0265} = 3770$$

and the improvement is 377 times.

3-11.　The answer is 18.5 kΩ.

3-12.　(a)　Both diodes are reverse-biased, so that

$$E_{out} = E_{in} = 1 \text{ V}$$

(The diodes have no effect in this case.)
　　　(b)　Since (-2.0) is more positive than is (-10), diode D_2 is forward-biased; the voltage across it becomes about -0.5 V. Therefore, the output is independent of the input and is

$$E_{out} = -2.0 - 0.5 = -2.5 \text{ V}$$

The drop across R is 10 - 2.5 = 7.5 V.

　　　(c)　By arguments similar to those in (b) one can see that the output is limited (clipped) to about ±2.5 V, giving an approximation to a square wave. (see the figure.)

　　　(d)　In this case, a ramp appears between +2.5 V and - 2.5 V, as in the figure.

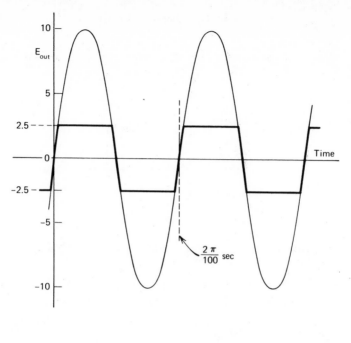

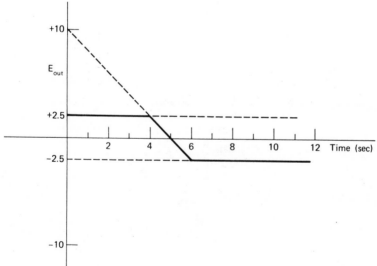

4-1. The computation can be carried out by a circuit such as shown in the figure.

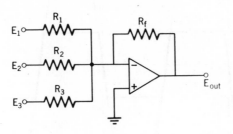

To satisfy the required coefficients, the following ratios must be met: $R_f/R_1 = 5$, $R_f/R_2 = 3$, $R_f/R_3 = 1$. In order to give convenient values for all resistors, a value of R_f should be selected which is evenly divisible by 5, 3, and 1, say 15 kΩ. Then $R_1 = 3$ kΩ, $R_2 = 5$ kΩ, and $R_3 = 15$ kΩ.

4-2. The equation is

$$E_{out} = - E_{in}$$

4-3. A suitable circuit is shown in the figure.

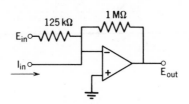

4-4. For Figure 5-6b, $E_{out} = -kE_{in}$

For c, $E_{out} = - \frac{1-k}{k} \cdot E_{in}$

For d, $E_{out} = -kRI_{in}$

Note that c, although it permits a wide variation in the gain, is far from linear.

4-5.

$$E_{out} = -\frac{E_{in}}{RC} \cdot t$$

$$= -\frac{(10^{-3})(5)(60)}{(6 \times 10^5)(0.5)(10^{-6})} = -1 \text{ V}$$

4-6.   It is convenient to establish an intermediate potential at point E in Figure 4-40.

$$E = \frac{-1}{(10^6)(10^{-6})} \int 2 \, dt = -2t \text{ and } \frac{dE}{dt} = -2$$

$$E_{out} = -(2 \times 10^6)(2 \times 10^{-6}) \frac{dE}{dt} = -4 \frac{dE}{dt}$$

Hence $E_{out} = +8$ V.   Note that E_{out} is time independent.

4-7.   See Figure.

(a)   When E_{in} is negative or zero,

$$E_{out} = -\frac{R_f}{R_{in}} \cdot E_{in}$$

When E_{in} is positive, the diode is reverse-biased, and $E_{out} = -10$ V (saturation).

(b)   When $E_{in} > 0$, $E_{out} = -10$ V. At $E_{in} = 0$, no current flows, and the voltage across the diode must be zero, so $E_{out} = +2$ V. When $E_{in} < 0$, the output remains at +2 V because there is no feedback resistor. Remember that the summing junction is at virtual ground.

(c)   When $E_{in} > 0$, $E_{out} = -10$ V. At $E_{in} = 0$, by the same argument as above, $E_{out} = -2$ V. At $E_{in} < 0$, the output is -2 V.

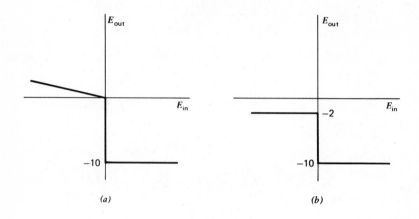

(a)

(b)

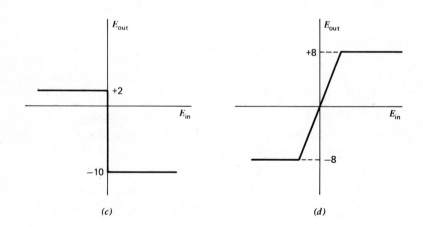

(c)

(d)

(d) $E_{out} = - (R_f/R_{in})E_{in}$, with bounds at ±8 V.

4-8. (a) When $E_{in} < 0$, the diode is reverse-biased, and the circuit is a simple voltage follower, $E_{out} = E_{in}$. When $E_{in} > 0$, the circuit becomes a follower-with-gain, and the gain increases exponentially due to the logarithmic characteristic of the diode. The shape of the curve depends on the value of R_f. (See the figure.)

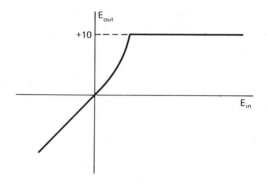

(b) See Figure 4-23b.
(c) This is a voltage-follower, and neither resistor has any effect on the output (for an ideal amplifier). $E_{out} = E_{in}$.

5-1. (a)
$$E_Q = - \frac{1}{RC} \int E_p \, dt$$

$$\frac{dE_Q}{dt} = - \frac{1}{RC} E_p$$

$$E_Q = E_p$$

Hence the general equation is

$$\frac{dx}{dt} = - \frac{1}{RC} x$$

Mathematically, the solution is

$$x = A \exp\left(-\frac{t}{RC}\right)$$

where A is the initial condition voltage.

(b)

$$E_Q = + \frac{1}{RC} \int E_p \, dt$$

$$\frac{dx}{dt} = + \frac{1}{RC} x$$

which has the mathematical solution

$$x = A \exp \frac{t}{RC}$$

5-2.   The first step in solving this type of problem is to assign to point P a derivative of an order equal to the number of integrations around the loop. Then the outputs of the successive amplifiers can be expressed in terms of various lesser derivatives.

(a) $E_p = \dfrac{dx}{dt}$ $E_{p'} = 2$ V

$$E_1 = - \frac{1}{RC} x \qquad E_2 = - \frac{2}{RC} t$$

$$E_3 = E_2 - E_1 = - \frac{2}{RC} t + \frac{1}{RC} x$$

Since in the given example $RC = 1$ sec,

$$E_3 = x - 2t$$

and since $E_3 = E_p$, we can write

$$\frac{dx}{dt} = x - 2t$$

which is the required equation.

(b)

$$E_P = \frac{d^3x}{dt^3} \qquad E_1 = - \frac{1}{RC} \cdot \frac{d^2x}{dt^2}$$

$$E_2 = \left(\frac{1}{RC}\right)^2 \frac{dx}{dt} \qquad E_3 = -\left(\frac{1}{RC}\right)^3 x$$

$$E_4 = -E_2 - E_{33} = -\left(\frac{1}{RC}\right)^2\left(\frac{dx}{dt}\right) + \left(\frac{1}{RC}\right)^3 x$$

Since $RC = 1$ and $E_4 = E_p$, we have

$$\frac{d^3x}{dt^3} = -\frac{dx}{dt} + x$$

which is the desired equation for (b).

(c)   The upper loop is considered to operate on the variable x, the lower loop on y. The value of E_2 is determined by analysis of the upper loop and is added, where appropriate, to the equation for the lower loop. The result consists of a pair of simultaneous second-order differential equations.

$$E_p = \frac{d^2x}{dt^2} \qquad E_{p'} = \frac{d^2y}{dt^2}$$

$$E_1 = -\left(\frac{1}{RC}\right)\left(\frac{dx}{dt}\right) \qquad E_4 = -\left(\frac{1}{RC}\right)\left(\frac{dy}{dt}\right)$$

$$E_2 = \left(\frac{1}{RC}\right)^2 x - \qquad E_5 = \left(\frac{1}{RC}\right)^2 y$$

$$E_6 = -\left(\frac{1}{RC}\right)^2 x - \left(\frac{1}{RC}\right)^2 y \qquad E_3 = -\left(\frac{1}{RC}\right)^2 x$$

Since $RC = 1$,

$$E_6 = -(x + y) \qquad\qquad E_3 = -x$$

and

$$\begin{cases} \dfrac{d^2x}{dt^2} = -x \\[2ex] \dfrac{d^2y}{dt^2} = -(x + y) \end{cases}$$

is the required solution.

5-3.   In b of the figure note that since $E_2 = x$,

$$6 - 0.531 \; \frac{10^6}{(5)(10^4)} \quad E_2 = 6 - 10.62x$$

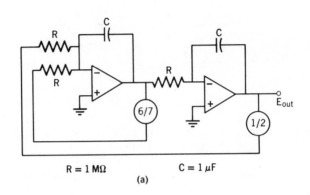

R = 1 MΩ   C = 1 μF

(a)

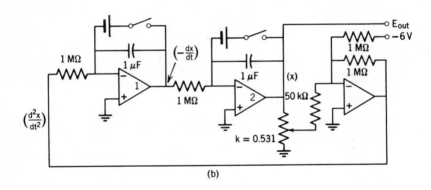

(b)

5-4. The general solution is $x = A \sin (\omega t + \phi)$

and $$\frac{dx}{dt} = A \omega \cos (\omega t + \phi)$$

For each case, substitute 0 for t (initial conditions refer to time $t = 0$) and note that the sine function is zero for zero angle, while the cosine has its maximum amplitude at zero angle.

(a) $A \omega \cos (0 + \phi) = 0$

            $A \sin (0 + \phi) = 2$

Consequently, $A = 2$ V, $\phi = 90°$.

(b) $A\omega \cos (0 + \phi) = 2$

$A \sin (0 + \phi) = 0$

from which, $A = (2/\omega)$ V, $\phi = 0°$. (This is inconvenient for amplitude control of a sine wave, since A is a function of ω.)

(c) $A\omega \cos (0 + \phi) = 2$

$A \sin (0 + \phi) = 2$

In this case, one can write

$$\frac{\sin \phi}{\cos \phi} = \frac{A\omega}{A} = \omega = \tan \phi$$

so that for every ω there are values for ϕ and A.

(d) $A\omega \cos (0 + \phi) = 0$

$A \sin (0 + \phi) = 0$

These two equations are compatible only when $A = 0$, hence ϕ is not defined.

5-5. In the lower loop, variation of R_Q changes the frequency. When the attenuation at R_Q is zero, the frequencies generated in the two loops are equal, and $E_4 = 0$ (if the phases and amplitudes are identical). If either differs a sine wave of the same frequency will appear. As the attenuation at R_Q is increased, the frequency of the lower loop decreases, and a mixed signal appears at the output:

$$-A \sin (\omega_1 t + \phi_1) + B \sin (\omega_2 t + \phi_2)$$

When attenuation is complete, P' is grounded, and the B signal disappears entirely. The output remains at $-A \sin (\omega_1 t + \phi_1)$.

6-1. The drop across R will be $15.0 - 12.0 = 3.0$ V. The maximum load current, $I_{L(max)}$, is

$$I_{L\,(max)} = \frac{12.0}{100} = 0.120 \text{ A}$$

The minimum is $I_{L\,(min)}$, given by

$$I_{L\,(min)} = \frac{12.0}{2000} = 0.006 \text{ A} = 6 \text{ mA}$$

At maximum load current, the zener current is a minimum. Let

$$I_{z\,(min)} = 0.010 \text{ A}$$

At minimum load current, the zener current is a maximum:

$$I_{z\,(max)} = I_{z\,(min)} + I_{L\,(max)} - I_{L\,(min)}$$

$$= 0.010 + 0.120 - 0.006 = 0.124 \text{ A}$$

The maximum zener dissipation is $I_{z\,(max)} E_z$

$$= (0.124)(12) = 1.49 \text{ W}$$
$$\text{(use a 2-W unit)}$$

The value of R must be

$$R = \frac{3.0}{I_{z\,(min)} + I_{L\,(max)}} = \frac{3.0}{0.13} = 23 \ \Omega$$

The power rating of the resistor must be $P = EI$

$$P = (3)(0.13) = 0.39 \text{ W (use a } \tfrac{1}{2} \text{ W resistor}$$
$$\text{or larger)}$$

6-2. The current through R is given by

$$I_R = \frac{20 - 10.5}{100} = 0.095 \text{ A}$$

(It is assumed that $V_{BE} = 0.5$ V.)
The power rating then is

$$P_R = I_R V_R = (0.095)(9.5) = 0.90 \text{ W}$$

The maximum transistor current occurs when the power supply is not loaded, and the power dissipation of the transistor-diode pair is

$$P = (0.095)(10.5) = 1 \text{ W}$$

Between these two components the dissipation is distributed in the ratio $P_T/P_z \cong \beta$, a large number; consequently, the power requirement of the zener is negligible.

6-3. In the ON condition (when E_{in} is positive), E_{out} is

$$E_{out} = \frac{1000}{1001} E_{in} = 0.999\, E_{in}$$

When the diode is OFF, E_{out} becomes

$$E_{out} = \frac{1000}{1,001,000} E_{in} = 0.001\, E_{in}$$

6-4. (a)
$$I_B = \frac{1.0 - V_{BE}}{10000} \cong \frac{1.0}{10^4} = 10^{-4}\ \mathrm{A} = 0.1\ \mathrm{mA}$$

(b) $I_C = (\beta - 1)I_B \cong 20.0\ \mathrm{mA}$

(c) $E_{out} = 25 - 1000 I_C = 25 - 20 = 5\ \mathrm{V}$

(d) Assume that a 1 mV signal is applied, which produces

$$I_B = 0.1\ \mu\mathrm{A}, \text{ hence } I_C = 20\ \mu\mathrm{A}, \text{ hence}$$

$$E_{out} = 20\ \mathrm{mV}. \quad \text{From these data}$$

$$A_v = \frac{20}{1} = 20$$

6-5. (a)
$$I_C = (\beta - 1)I_B \cong (200)(10^{-5})$$

$$= 2 \times 10^{-3}\ \mathrm{A} = 2\ \mathrm{mA}$$

$$E_{out} = 25 - (0.002)(1000) = 23\ \mathrm{V}$$

(b) $\dfrac{dE_{out}}{dI_{in}} \cong \dfrac{\Delta E_{out}}{\Delta I_{in}}$

Choose $\Delta I_{in} = 1\ \mu\mathrm{A}$, hence $\Delta E_{out} = (200 \times 10^{-6}) \times (10^3)\ \mathrm{V}$

Hence

$$\frac{dE_{out}}{dI_{in}} = \frac{200 \times 10^{-3}}{10^{-6}} = 2 \times 10^5 = 0.2\ \mathrm{m}\Omega$$

(c)
$$R_B = \frac{25 - V_{BE}}{10 \times 10^{-6}} \cong \frac{25}{10^{-5}} = 25 \times 10^5$$
$$= 2.5 \text{ M}\Omega$$

7-1. See the figure. In b, the average output is zero if the zero-crossings of the signal and the chopper coincide. Any second harmonic (100 Hz) present in the 50-Hz wave will *not* be averaged out and will appear at the output. The circuit can serve as an harmonic distortion sensor.

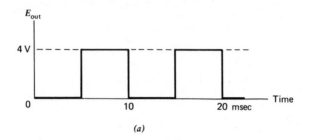

(a)

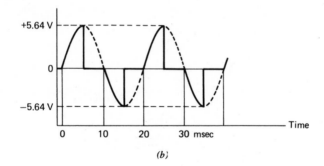

(b)

7-2. The combination of C_1 and C_2 has a capacitance of
$$C_T = \frac{C_1 C_2}{C_1 + C_2}$$

Therefore the frequency becomes
$$f = \frac{1}{2\pi\sqrt{LC_1 C_2 / (C_1 + C_2)}}$$

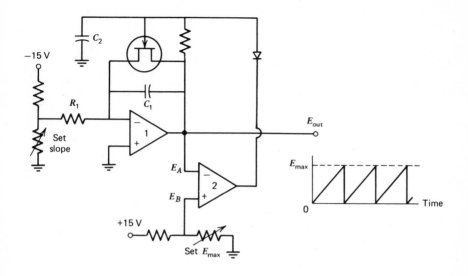

7-3. The integrator (see Figure) produces a positively
      going ramp. As long as amplifier 2 (a compara-
      tor) senses $(E_B - E_A) > 0$, its output will be
      -10 V and the FET will be off. When the ramp
      reaches E_{max}, the comparator shifts to $+10$ V, and
      the FET goes into conduction, discharging C_1.
      Capacitor C_2 may be needed to lengthen the dis-
      charge time of C_1 sufficiently to ensure that the
      voltage reaches zero. Note that in this circuit
      the frequency is not directly controllable, only
      E_{max} and the slope are accessible.

8-1.

A	B	E_{out}
0	0	1
0	1	0
1	0	1
1	1	0

(a)

A	B	C	E_{out}
0	0	0	0
0	0	1	1
0	1	0	1
0	1	1	1
1	0	0	0
1	0	1	1
1	1	0	0
1	1	1	0

(b)

A	B	E_{out}
0	0	1
0	1	0
1	0	0
1	1	1
(c)		

This is sometimes called an *equality-comparator* or *co-incidence* circuit.

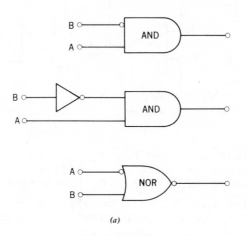

(a)

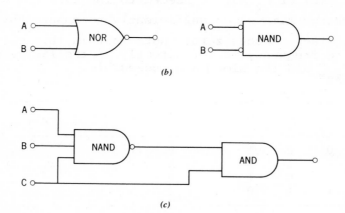

(b)

(c)

8-2.   (a)   Three equivalent forms are shown in the Fig-
             ure.
       (b)   Two equivalent forms (see the figure).

             In c the NAND gate always gives "1," so that
             the output is equal to the C input, except
             for the last case, where all inputs are "1."

8-3.

A	B	Carry	Sum
0	0	0	0
0	1	0	1
1	0	0	1
1	1	1	0

The circuit is called a "half-adder." It performs the addition operation with a "carry" output to permit a multibit addition. (The circuit
is available as an IC.)

8-4.   The circuit is shown in the diagram. In the Gray
       code two consecutive numbers differ in only *one*
       of the four bits, ABCD, as the reader can ascer-
       tain. It is widely used in computer encoding.

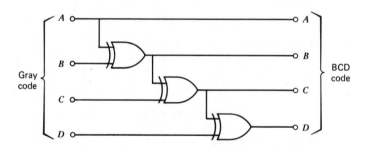

8-5.

A	B	C	D
0	0	0	0
0	0	1	1
0	1	0	1
0	1	1	0
1	0	0	1
1	0	1	0
1	1	0	0
1	1	1	1

Note that D is 1 if the number
of 1's at the input is odd and
0 if there is an even number of
1's. This circuit, called a
parity check, is used to identify errors, since most errors
change only one bit, and hence
cause a change in the number of
1's. (On the back cover of most

recent books one can find an identifying number
called the International Standard Book Number
(ISBN). The last symbol is a special version of
a parity check.)

8-6. See the Figure.

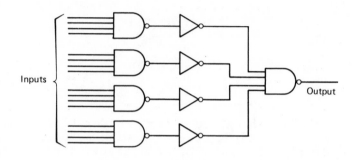

8-7. (a) A binary number can be expressed as

$$(0 \text{ or } 1)2^4 + (0 \text{ or } 1)2^3 + (0 \text{ or } 1)2^2$$
$$+ (0 \text{ or } 1)2^1 + (0 \text{ or } 1)2^0$$

in which each term represents one digit.
Consequently the numbers called for are:

(a)   01001 = 0 + 8 + 0 + 0 + 1 = 9
(b)   11000 =16 + 8 + 0 + 0 + 0 = 24
(c)   10001 =16 + 0 + 0 + 0 + 1 = 17
(d)   11111 =16 + 8 + 4 + 2 + 1 = 31

(b) In order to obtain a binary transformation,
divide the number successively by decreasing powers of 2: Thus to convert 17 to binary,

(e)   17 ÷ 16 = 1, remainder 1
         1 ÷ 8 = 0, remainder 1
         1 ÷ 4 = 0, remainder 1
         1 ÷ 2 = 0, remainder 1
         1 ÷ 1 = 1, remainder 0

Hence, 17 = 10001

By this method:

(f) 168 ÷ 128 = 1, remainder 40
      40 ÷ 64 = 0, remainder 40
      40 ÷ 32 = 1, remainder 8
       8 ÷ 16 = 0, remainder 8
       8 ÷   8 = 1, remainder 0
       0 ÷   4 = 0,
       0 ÷   2 = 0,
       0 ÷   1 = 0

Hence, 168 = 10101000

Similarly,

(g)   4 = 100
(h) 44 = 101100

9-1.

	R	S	Q	Q̄
(1)	1	1	0	0
(2)	0	1	1	0
(3)	1	0	0	1
(4)	0	1→ 0	1	0
(5)	1→ 0	0	0	1

The above truth table applies when the clock input, C, is "1." When $C = 0$, case (1) is indeterminate, and in all other cases, the previous logic state is maintained.

9-2. Notice that this circuit is similar to that of Figure 9-3, with one inverting input and with R and S inputs connected together. Hence only lines 2 and 3 of the truth table remain active. The output at Q follows the input for $C = 1$ and stores this information when $C = 0$, regardless of changes at the input. This may be called a sample-and-hold operation.

9-3. The collection of flip-flops has an inherent count of 64. The AND gate responds to 111100 = 60, and resets the flip-flops. Hence this circuit constitutes a generator of short-duration pulses repeating at the rate of 1 Hz.

9-4. The logic diagram for this figure is identical with that shown in Figure 9-3.

10-1.   Let $x = -t/RC$, $f = 1 - e^x$

Then

$$f = 1 - e^x = 1 - (1 + \frac{x}{1} + \frac{x^2}{2} + \frac{x^3}{6} + \ldots)$$

$$= x + \frac{x^2}{2} + \frac{x^3}{6} + \ldots$$

The linear approximation is $f \cong x$, while most of the error comes from the term in x^2. So neglect the terms in x^3 and higher. The 10% error limit is reached when $x^2/2 = 0.10x$ or $x = 0.20$. Consequently, less than 10% error will be encountered if the function $f \cong -t/RC$ is used instead of the actual function, $f = 1 - e^{-t/RC}$ for t/RC < 0.20. In time units, this corresponds to the time elapsed between zero and $0.20RC$ second.

10-2.   The governing equation is $C = 0.0885\ \varepsilon a/d$. If the variable of interest is denoted by x, then $a = k\sqrt{x}$, or $x = a^2/k^2$, where k is a constant.

$$C = 0.0885 \left[\frac{\varepsilon}{d}\right] k\sqrt{x}$$

$$x = \left[\frac{d}{0.0885\ \varepsilon\ k}\right]^2 C^2$$

which is the desired function.

10-3.

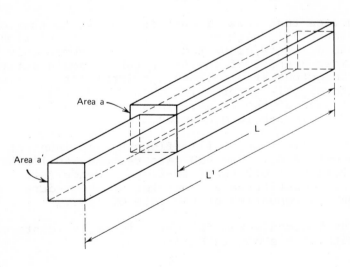

The volume of the piece remains the same, but the length increases from L to L', and the cross-sectional area diminishes from a to a', as shown in the sketch. Hence $La = L'a'$, or $a/a' = L'/L$. The corresponding resistances are R and R'.

$$R' = R\left(\frac{L'}{L}\right)\left(\frac{a}{a'}\right) = R\left(\frac{L'}{L}\right)^2$$

$$R = \rho L a^{-1} \quad \text{(where } \rho \text{ is the resistivity)}$$

$$R' = \rho L a^{-1}\left(\frac{L'}{L}\right)^2$$

$$\Delta R = R' - R = \rho L a^{-1}\left(\frac{L'}{L}\right)^2 - \rho L a^{-1}$$

$$= \rho L a^{-1}\left[\left(\frac{L'}{L}\right)^2 - 1\right]$$

$$\Delta L = L' - L$$

Hence

$$S = \frac{L \; \Delta R}{R \; \Delta L} = \frac{L \rho L a^{-1}[(L'/L)^2 - 1]}{\rho L a^{-1}(L' - L)}$$

which simplifies algebraically to

$$S = \frac{L' + L}{L}$$

Since L' is only very slightly greater than L, this expression is almost exactly 2.

10-4.   First, compute the circuit resistance at 350°K:

$$R_{350°} = 10[1 + (0.01)(50)] = 15 \ \Omega$$

The total resistance then is

$$R_T = 100 + 15 = 115 \ \Omega$$

The voltage is $(800 - 300)(2 \times 10^{-6}) = 10^{-3}$ V $= 1$ mV.
Hence the meter current is

$$\frac{1 \text{ mV}}{115 \ \Omega} = \frac{10^{-3}}{115} = 8.7 \ \mu A$$

10-5. Compute first the minimum output voltage and input current:

$$I_{in} = \frac{10 \text{ nA}}{0.05} = 200 \text{ nA} = 0.2 \text{ }\mu\text{A (minimum)}$$

$$E_{out} = \frac{2 \text{ mV}}{0.05} = 40 \text{ mV (minimum)}$$

We can calculate

$$R_f = \frac{40 \text{ mV}}{0.2 \text{ }\mu\text{A}} = 0.20 \text{ M}\Omega$$

as the minimum value for R_f; lower values will give insufficient voltage, higher values will not significantly improve the S/N ratio.

The current, 0.20 µA corresponds to

$$\frac{(2)(10^{-7})}{(1.6)(10^{-19})} = (1.25)(10^{12}) \text{ electrons per second,}$$

which in turn is produced by

$$\frac{(1.25)(10^{12})}{(10^7)(0.20)} = (6.25)(10^5) \text{ photons per second.}$$

10-6.

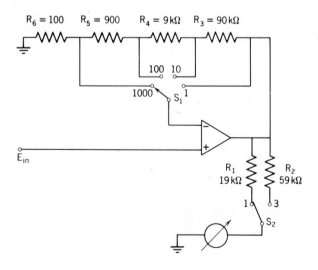

The switching arrangement shown in the figure is one of many that will give the desired results. The two switches can easily be combined in one double-deck rotary switch to give the desired sequence. The individual resistors on S_2 are calculated as follows:

$$R_1 + R_M = \frac{1}{(50)(10^{-6})} = 20 \text{ k}\Omega$$

so that $R_1 = 19 \text{ k}\Omega$

$$R_2 + R_M = 3(R_1 + R_M) = 60 \text{ k}\Omega$$

so that $R_2 = 59 \text{ k}\Omega$

With switch S_1 in position 1, the amplifier acts as a simple voltage follower and, depending on S_2, reads either 1 or 3 V FS. In positions 10, 100, and 1000 it becomes a follower-with-gain and gives successively higher sensitivity.

The total resistance of the feedback divider is arbitrary; a suitable value would be 100 kΩ. In this case, for the switch in position 10:

$$\frac{R_4 + R_5 + R_6}{100,000} = 0.1$$

$$R_4 + R_5 + R_6 = 10,000 \text{ so that } R_3 = 90 \text{ k}\Omega$$

For position 100:

$$\frac{R_5 + R_6}{100,000} = 0.01$$

$$R_5 + R_6 = 1000, \text{ so that } R_4 = 9 \text{ k}\Omega$$

For position 1000:

$$\frac{R_6}{100,000} = 0.001, \text{ so that } R_6 = 100 \text{ k}\Omega$$

and $R_5 = 900 \text{ k}\Omega$

11-1. (a, b)   Identical diagrams. (See the figure).
           The diode is shown in the conducting posi-
           tion, that is, with the switch "open," in
           that no signal will be seen at the output.

       (c)   The diagram here is nearly identical with
           Figure 8-3.

       (d)   (See the figure.)   The four switches are
           considered to be connected mechanically so

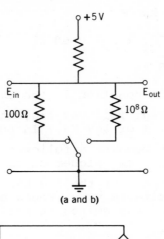

(a and b)

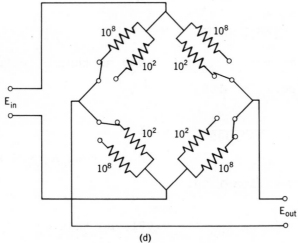

(d)

that they all move together from the positions
shown to the opposite poles.

11.2   (See the figures.)

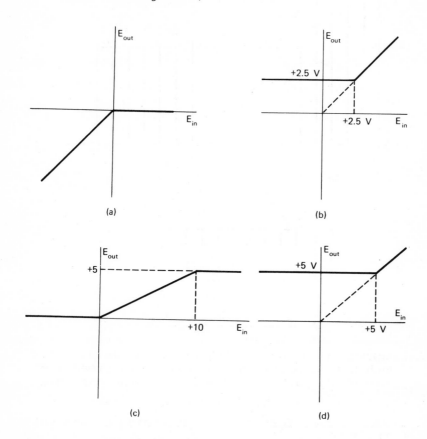

(a)

(b)

(c)

(d)

11-3.   (See the figures.)

11-4.   (a)   Identical to 11-3*b*.

        (b)   The same, except from -6 to 0 V.

        (c)   See the figure.

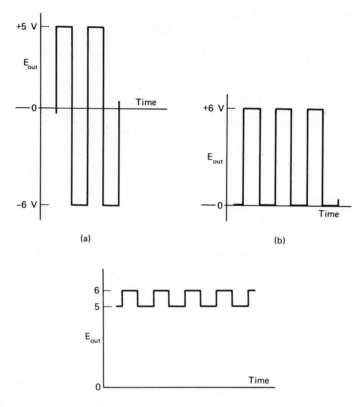

11-5. Depending on the setting of K, the output of
each potentiometer has a voltage intermediate
between E_{in} and + 1 or - 1 V, as the case may be.
As E_{in} varies, this voltage also varies, and
crosses zero for some E_{in} depending on the setting. Beyond this point the diode is forward
biased, and the effect is of switching in the
corresponding input resistor.

(a) For potentiometer 1, $K_1 = \frac{1}{2}$, and the voltage is halfway between E_{in} and +1 V. Hence the
breakpoint, where the diode becomes conductive,
is at E_{in} = -1 V (neglecting the forward drop of

the diode). At this point, one resistor is effectively switched in, and the voltage produced by the potentiometer is

$$E_{in} + k_1(1 - E_{in})$$

The output is

$$E_{out} = -\frac{R_f}{R}[E_{in} + k_1(1 - E_{in})]$$

$$= -\frac{R_f}{R}[E_{in}(1 - K_1) + K_1]$$

Since $k_1 = \frac{1}{2}$, we can write

$$E_{out} = -(\tfrac{1}{2}E_{in} + \tfrac{1}{2}) \quad \text{(Valid for } E_{in} < -1 \text{ V.)}$$

This corresponds to the B segment of the curve. Similarly, the break point for segment A is at the point where D_2 becomes forward biased, and

$$E_{in} + k_2(1 - E_{in}) = 0$$

For $k_2 = \frac{3}{4}$, E_{in} turns out to be -3 V. At this

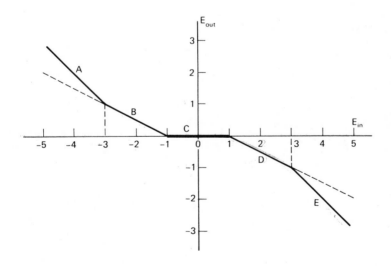

point, a second resistor R is switched in parallel to the first, so that the slope of the curve increases abruptly. Segment C shows zero output because no diode is conducting. Segments D and E can be calculated similarly. The complete curve is shown in the figure.

(b)   Changing k varies both the break voltage and the slope, as shown by the equation

$$E_{out} = - \frac{R_f}{R} [E_{in} (1 - k) + k]$$

(c)   Varying R changes the slope, but does not affect the break point.

Note that by using a large number of diode branches, it is possible to approximate closely any single-valued function passing through the origin.

11-6.   (See the figures.)

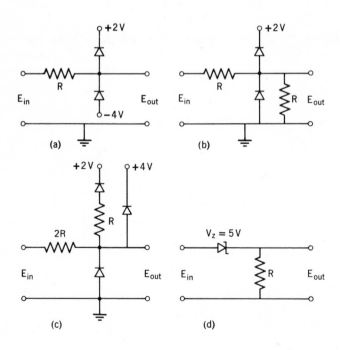

(a)   (b)   (c)   (d)

12-1. (a) $(27 + j0)$

(b) $(2 + j2)$

(c) $+j18$

(d) $(3 - j4)$

12-2. (a) $(8\ \underline{/37°})$

(b) $(2\ \underline{/97°})$

(c) $j31 - (3\sqrt{2} + j3\sqrt{2}) = [-3\sqrt{2} + j(31 - 3\sqrt{2})]$

(d)   By extending the procedure of multiplication, the amplitude is raised to the 2.5 power, while the angle is multiplied by 2.5. Hence the answer is $A^{2 \cdot 5}\ \underline{/2.5\theta}$.

12-3. (a) $(6\sqrt{2}\ \underline{/225°})$

(b) $|W| = \sqrt{x^2 + y^2} = \sqrt{16 + 48} = 8$

$\theta = \arctan \dfrac{-4\sqrt{3}}{4} = \arctan(-\sqrt{3}) = 300°$

Answer: $(8\ \underline{/300°}) = (8\ \underline{/-60°})$

(c) $x = |W| \cos\theta = 11\ \dfrac{1}{2} = 5.5$

$y = |W| \sin\theta = 11\ \dfrac{\sqrt{3}}{2} = 5.5\sqrt{3}$

Answer: $(5.5 + j5.5\sqrt{3})$

(d) $-j11$

12-4. (a) $Z = \dfrac{j\omega LR}{j\omega L + R}$

(b) $\dfrac{1}{Z} = \dfrac{1}{R} + \dfrac{1}{R + 1/(j\omega C)} = \dfrac{R + (R + 1/j\omega C)}{R(R + 1/j\omega C)}$

$Z = \dfrac{R^2 + R/j\omega C}{2R + 1/j\omega C} = \dfrac{j\omega R^2 C + R}{2j\omega RC + 1}$

(c) $\dfrac{1}{Z} = \dfrac{1}{R} + \dfrac{1}{j\omega L} + j\omega C = \dfrac{j\omega L + R + j^2\omega^2 LCR}{j\omega RL}$

$Z = \dfrac{j\omega RL}{j\omega L + R - \omega^2 LCR}$

(d) $\dfrac{1}{Z} = \dfrac{1}{j\omega L} + j\omega C = \dfrac{1 + j^2\omega^2 LC}{j\omega L}$

$Z = \dfrac{j\omega L}{1 - \omega^2 LC}$

12-5. To operate normally, the lamp must carry 1 A (RMS) (12 W/12 V). Its impedance = 12 Ω.

(a) $Z_T = Z_C + Z_{bulb} = Z_C + 1$

$Z_C = \dfrac{1}{j2\pi 60 C}$

$I = \dfrac{115}{1/j120\pi C + 12} = 1$ A

Solve this for C. Since the phase angle introduced by the capacitor does not affect the lamp, we can write

$Z = \left| \dfrac{1}{j120\pi C} + 12 \right|$ which must equal 115 Ω.

Take the absolute value

$$\sqrt{\left(12 + \dfrac{1}{j120\pi C}\right)\left(12 - \dfrac{1}{j120\pi C}\right)} =$$

$$\sqrt{12^2 + \left(\dfrac{1}{120\pi C}\right)^2} = 115$$

from which it follows that

$\left(\dfrac{1}{120\pi C}\right)^2 = (115)^2 - 12^2$ (Ignore the 12^2.)

$C = \dfrac{1}{(120\pi)(115)} = 23$ μF

(b) The bulb has a 12 V (RMS) drop. The phase angle is arctan $(1/120\pi C)$.
Since $1/120\pi C$ is very large, the angle $\cong$ 90°. Thus the voltage drop across C (maximum) = (115)(1.41) V. Hence the voltage rating of the capacitor must be > 162 V. (In practice, an ample safety factor is always included.)

(c)   The power dissipated in the capacitor is EI:

$$EI = (E_C \underline{/0°})(I_C \underline{/90°}) = (E_C + j0)(0 + jI_C)$$

$$= 0 + jE_C I_C$$

The real part being zero, the power dissipated is also zero.

12-6.   (a) $$\frac{E_{out}}{E_{in}} = -\frac{Z_f}{Z_{in}} = -\frac{1/[(1/R_2) + j\omega C]}{R_1}$$

$$= -\frac{R_2}{R_1} \cdot \frac{1}{1 + j\omega R_2 C}$$

(b) $$\frac{E_{out}}{E_{in}} = -\frac{Z_f}{Z_{in}} = -\frac{1/[(1/R) + j\omega C_2]}{1/j\omega C}$$

$$= -\frac{j\omega R C_1}{j\omega R C_2 + 1}$$

(c)   Each impedance is equal to $2R + 2/j\omega C$; hence the transfer coefficient is unity.

12-7. $E = A_0 + A_1 \cos\theta + A_2 \cos 2\theta + \ldots$

$\qquad\qquad + B_1 \sin\theta + B_2 \sin 2\theta + \ldots$

$$A_0 = \frac{1}{\pi}\int_{-\pi}^{\pi} F(\theta)\, d\theta = \frac{1}{\pi}\int_{-\pi}^{0} -E_0\, d\theta + \frac{1}{\pi}\int_{0}^{\pi} E_0\, d\theta$$

$$= -\frac{1}{\pi} E_0 \theta \Big|_{-\pi}^{0} + \frac{1}{\pi} E_0 \theta \Big|_{0}^{\pi} = E_0 - E_0 = 0$$

(Hence the DC term is zero.)

$$A_n = \frac{1}{\pi}\int_{-\pi}^{0} F(\theta) \cos n\theta\, d\theta + \frac{1}{\pi}\int_{0}^{\pi} F(\theta) \cos n\theta\, d\theta$$

$$= -\frac{E_0}{\pi} \int_{-\pi}^{0} \cos n\theta \; d\theta \; + \; \frac{E_0}{\pi} \int_{0}^{\pi} \cos n\theta \; d\theta$$

$$= -\frac{E_0}{n\pi} \int_{n\theta=-n\pi}^{0} \cos n\theta \; dn\theta \; + \; \frac{E_0}{n\pi} \int_{0}^{n\theta=n\pi} \cos n\theta \; dn\theta$$

$$= -\frac{E_0}{n^2\pi} \sin n\theta \; \Bigg|_{n\theta=-n\pi}^{0} \; + \; \frac{E_0}{n^2\pi} \sin n\theta \; \Bigg|_{0}^{n\theta=n\pi}$$

Evaluation between the indicated limits gives zero for both terms; therefore there are no cosine terms in the expansion. (See the figure)

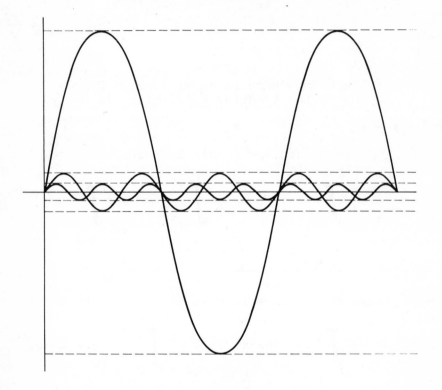

By a similar procedure, one finds

$$B_n = + \left.\frac{E_0}{n\pi}\cos n\theta\right|_{n\theta=-n\pi}^{0} - \left.\frac{E_0}{n\pi}\cos n\theta\right|_{0}^{n\theta=n\pi}$$

Remembering that $\cos 0 = 1$ and $\cos \pi = -1$, one can state that

for $n = 1, 3, 5, \ldots$, the terms add
for $n = 2, 4, 6, \ldots$, the terms cancel

Hence $B_n = 4E_0/\pi n$ for odd values of n.
The series thus is

$$E = \frac{4E_0}{\pi}\left[\sin\theta + \frac{1}{3}\sin 3\theta + \ldots\right]$$

12-8. $\theta = 2\pi ft; \qquad f = \frac{1}{T}$

For $t = -T/2$, write $\theta = 2\pi(1/T)(-T/2) = -\pi$
For $t = +T/2$, write $\theta = 2\pi(1/T)(+T/2) = +\pi$

12-9.   Note that the transfer coefficient $1/j\omega C$ is an impedance, so that the output must be a current, $I = Ej\omega C$. Consider the three terms separately:

First term, $\omega = 0$ (the DC term)

$I = 0$

Second term, $\omega = \omega_1$

$$I = (\frac{1}{3}\sin\omega_1 t)(j\omega_1 C) = j\omega_1 C(\frac{1}{3} + j0)$$

$$= \frac{1}{3}j\omega_1 C = \frac{1}{3}\omega_1 C\underline{/90°}$$

(This is a cosine wave.)

Third term, $\omega = 2\omega_1$

$$I = (\frac{1}{9}\sin 2\omega_1 t)(j2\omega_1 C) = j2\omega_1 C(\frac{1}{9} + j0)$$

$$= \frac{2}{9}j\omega_1 C = \frac{2}{9}\omega_1 C\underline{/90°}$$

(This is the second-harmonic cosine wave.)

Reverting to the trigonometric form,

$$I = 0 + \frac{1}{3} \omega_1 C \cos (\omega_1 t) + \frac{2}{9} \omega_1 C \cos (2\omega_1 t)$$

12-10. (a) Write the sum of voltage drops in terms of the current:

$$E_T = E_R + E_C + E_L$$

$$= IR + \cfrac{1}{\cfrac{1}{C} \int_0^t I \, dt} + L \frac{dI}{dt}$$

(b) Write the sum of currents in terms of the common voltage:

$$I_T = I_R + I_C + I_E$$

$$= \frac{E}{R} + C \frac{dE}{dt} + \cfrac{1}{L \int_0^t E \, dt}$$

12-11. (a) $\overline{E}_T = \overline{I} (R + \frac{1}{Cs} + Ls)$

(b) $\overline{I}_T = \overline{E} (\frac{1}{R} + Cs + \frac{1}{Ls})$

12-12. The desired equation is

$$Z_T = R + sL$$

For conditions just after time zero,

$$\frac{E_1}{s} = \overline{I} (R + sL)$$

$$\overline{I} = \frac{E_1}{s} \cdot \frac{1}{R + sL}$$

$$\overline{I} = \frac{E_1}{s} \cdot \frac{1}{s} \cdot \frac{1}{(R/L) + s}$$

By setting $R/L = a$, this can be rewritten as

$$\overline{I} = \frac{E_1}{L} \left[\left(\frac{1}{s}\right) \left(\frac{1}{a+s}\right) \right]$$

The Laplace table gives the inverse transform:

$$\frac{1}{s(s+a)} = \frac{1}{a} [1 - \exp(-at)]$$

Thus for the present case

$$I = \frac{E_1}{L} \cdot \frac{L}{R} \left[1 - \exp\left(-\frac{R}{L}t\right) \right]$$

$$I = \frac{E_1}{R} \left[1 - \exp\left(-\frac{R}{L}t\right) \right]$$

Hence the current starts at zero at $t = 0$ [since exp (0) = 1], and increases to E_1/R for large values of t.

INTRODUCTION

The following experiments are designed to accompany study of the text. They are divided into two sections. The first section consists of basic exercizes with directions spelled out in some detail. The experiments of the second group are more advanced; in some cases a circuit is presented only from a literature reference requiring some library work, and the student is asked to assemble the circuit and test it out.

The availability of a few standard items of test equipment is assumed. A digital multimeter ($3\frac{1}{2}$-digit accuracy) and an oscilloscope are essential. By far the best way to construct experimental circuits is on a prototyping board with built-in power supplies to give +5, +15, and -15 V of regulated DC. These are available from E & L Instruments, Inc., 61 First St., Derby, Connecticut 06418, and Continental Specialties Co., 44 Kendall St., New Haven, Connecticut 06511.

EXPERIMENT I

Resistor Networks

<u>Objective</u>: To examine typical small resistors and to verify the laws governing their combinations.

<u>Reference</u>: Chapter III.

<u>Theory</u>:   The governing equations are:
(1)   Ohm's law: $E = RI$
(2)   The power law: $P = EI = E^2/R = I^2R$

(3)   Parallel combination: $R_T = \dfrac{R_1 R_2}{R_1 + R_2}$

(4)   Series combination: $R_T = R_1 + R_2$

Procedure

(1)   Assemble the combination of components shown in Figure 1, where V is a voltmeter, 0 to 10 V, at least 1000 Ω/V; A is a milliammeter, 0 to 50 mA; E is a 9-V radio battery or equivalent. The resistor or combination of resistors to be tested is to be connected to the terminals marked X and Y. The fuse should be one that will burn out at 1/16A (60 mA); this is to protect both the battery and the ammeter from possible short circuits. (If digital meters are available, they can be used to advantage in this circuit.)
(2)   Obtain from the instructor three carbon or composition resistors with color-coding bands. Resist-

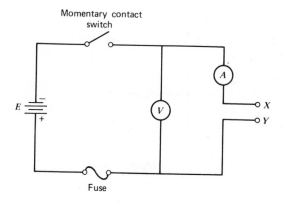

Figure 1.

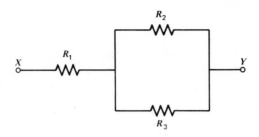

Figure 2.

ance values from a few hundreds to a few tens of thousands of ohms are appropriate. Determine from the col-
or code the nominal resistance of each and its tolerance rating (see Appendix VI). Connect each resistor
separately to the X and Y contacts, record the corresponding voltage and current readings, and compute the
resistance. How well do these values agree with the
ratings?
     (3) Connect the three resistors in series to X
and Y. How does the measured resistance compare with
the calculated value?

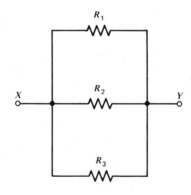

Figure 3.

(4)   Connect the three resistors in the network of Figure 2, and compare the measured and computed net resistances.

(5)   Repeat with the connection of Figure 3.

(6)   For each resistor in each of the combinations, calculate the power dissipated according to both versions of the power equation. Do they agree? How do they compare with the known power ratings of the resistors, as given you by the instructor? What would be the maximum voltage that it would be safe to apply to each network of resistors without exceeding the power rating of each individual resistor?

EXPERIMENT 2

Capacitors

Objective: To study the laws governing the combinations of capacitors.

Reference: Chapter III.

Theory: Capacitors follow the equations:

(1)   Definition: $I = C(dE/dt)$

(2)   Parallel combination: $C_T = C_1 + C_2$

(3)   Series combination: $C_T = \dfrac{C_1 C_2}{C_1 + C_2}$

Procedure

(1)   The circuit of Figure 4 is to be assembled. Calculate the maximum power that may be dissipated in each of the two resistors, and ask the instructor for properly rated resistors. E is a 9-V radio battery or an equivalent power supply. V is an electronic voltmeter with at least 10-MΩ input impedance.

(2)   To start the experiment, open switch S_1, close S_2, and wait until the voltage drops to zero. Then open S_1, simultaneously starting a timer. Record

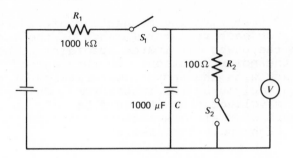

Figure 4.

the values of voltage every 10 seconds until it becomes
stable. (A strip-chart recorder can be used here to
advantage, if available.) Plot the voltage readings
versus time. Determine the RC time constant of the
circuit by observing the time needed for the capacitor
to charge to 63% of its final value.
      (3) Measure the value of R_1 (separated from the
circuit) with an ohmmeter. Calculate the value of C
and compare it with the labelled value. Comment about
the discrepance, if any. Estimate the precision of the
measurement in terms of possible errors in R_1, time,
and readout voltage. What merits do you see in this
method of measuring capacitance? Why did we insist
upon a high input-impedance voltmeter?

Additional Experiments

      Electrolytic capacitors have "memory" effects.
For example, when such a capacitor is discharged by
momentarily shorting its terminals, a lesser charge
will appear spontaneously on the capacitor. An enter-
prizing student can devise a suitable procedure to
measure this effect.

EXPERIMENT 3

AC Impedance

<u>Objective</u>: To illustrate the AC properties of capacitors and RC networks, and to learn to handle amplitude and phase relations.

<u>Reference</u>: Chapter III.

<u>Theory</u>: For more details of mathematical treatment, see Chapter XII.

Procedure

(1) Study the instructions for the oscilloscope to be used.

(2) Assemble the circuit of Figure 5a. The resistors can be small size($\frac{1}{4}$-W). Measure the amplitudes of the AC wave at points A and B, both relative to ground, by estimating the peak-to-peak distance on the oscilloscope screen. The value at B should be just half as great as that at A. If a two-trace scope is available, both signals may be displayed simultaneously, and relative amplitudes as well as phase differences, can be observed directly.

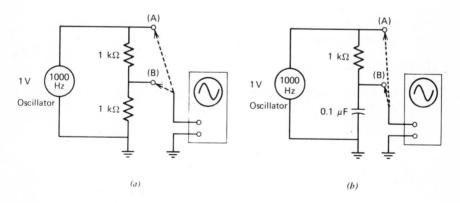

(a)   (b)

Figure 5. Dotted lines show alternative connections to the oscilloscope.

(3) Repeat the measurements of (2) using an AC
voltmeter in place of the oscilloscope. If the voltmeter is of the usual type, its measurements will be
RMS. Show how this reading can be made compatible with
the peak-to-peak readings from the scope.

(4) Now construct the circuit of Figure 5*b*. Again,
measure the voltages at *A* and *B* with respect to ground,
then measure the voltage appearing across the resistor
differentially, with the scope or meter connected directly to *A* and *B*, with no ground contact. Do the observed voltages across the resistor and capacitor add
up to give the total applied voltage? If not, explain
it.

(5) With the dual-trace scope, determine the
phase difference between the voltages appearing across
R and *C*. Calculate the sum of these two voltages in
complex notation (Chapter XII) and discuss your results.

Additional Experiments

Perform similar studies of circuits containing
inductors and resistors, and to circuits with all three
types of components included.

EXPERIMENT 4

RC-Filters

Objective: To illustrate the behavior of passive filters and to study component matching.

Reference: Chapter III.

Theory: *RC* filters are characterized by a critical frequency given by

$$f_0 = \frac{1}{2\pi RC}$$

The attenuation of a filter in decibels is given by the
formula

$$dB = 20 \log \frac{E_{out}}{E_{in}}$$

Procedure

(1)   Assemble a low-pass filter with oscillator and oscilloscope as in Figure 6*a*.   Select components R and C such that f_0 = 1000 Hz, approximately.   Keep in mind also that the impedances must be larger than about 1000 ohms to avoid loading the oscillator unduly, but less than about 100 kΩ to avoid making the circuit it-self sensitive to loading.   Note that at f_0, R and C have the same impedance.   There are many permissible combinations of R and C that will give the same criti-cal frequency, f_0.   Because of the difficulty in find-ing capacitors of precisely known values, the following procedure is suggested.   List several available resis-tors between 1000 Ω and 100 kΩ, and calculate for each the corresponding capacitance for the desired frequency. Select a capacitor from those available that is clos-est to one of your calculated values.   Keep in mind that capacitors usually have large tolerances of ±10 or 20% or even greater.

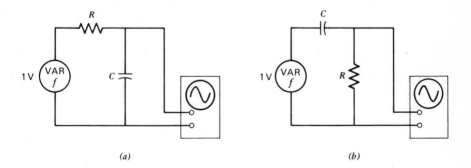

(a)   *(b)*

Figure 6.

   (2)   Test the frequency response of your filter by
determining the input and output amplitudes for frequencies.   Plot your results as a function of the logarithm of frequency, a Bode plot.
   (3)   Repeat the above with the high-pass circuit
of Figure 6b.
   (4)   With a dual-trace scope, determine the phase
difference between input and output for each filter at
various frequencies, and plot it against frequency.
Such plots are useful in predicting the dynamic behavior of filters.

Additional Experiments

   It is sometimes necessary to tune a filter to precisely the desired frequency.   This can most easily be
done by trimming the resistance values by adding low
value resistors in series, if R needs to be increased,
or large values in parallel if it needs to be lowered.
   Assume the desired correction in the RC product
to be $x\%$.   If RC is to be increased (positive x), add
a resistor in series with R, of value $R' = xR/100$
(Figure 7a).   To decrease RC, use a parallel resistor

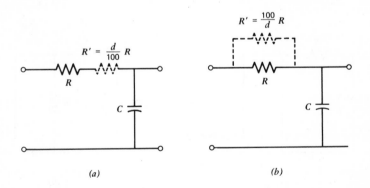

(a)   (b)

Figure 7.   Resistors R', shown in dashed lines, are for fine tuning.

$R' = 100R/x$ Ω (Figure 7b). Note that this last formula is only valid for corrections less than a few percent. In neither case does R' need to be of precision
better than ±10%.

Another area worth investigating is the effect of
the filter on square and triangular waves. For this
measurement, use a commercial generator of the appropriate waveforms. Note that for a low-pass filter,
beyond f_0 both square and triangular waves tend to become more like sine waves. Can you interpret this fact?

EXPERIMENT 5

Thevenin Equivalent Circuits

Objective: To illustrate the formation of a Thevenin
equivalent for purely resistive circuits.

Reference: Chapter III.

Theory: Consider any combination of voltage sources,
all of the same frequency, and of resistors, capacitors
and inductors. If any two points in this combination
are connected to the outside by two wires, the circuit
appears indistinguishable externally from a series combination of a single source E_{Th} and a single impedance
Z_{Th}. This is analagous to the observation that any
sum or product of mathematical fractions can be reduced to a single fraction.

Procedure

In this experiment we only use a DC source and
impedances consisting of resistors. Construct the circuit of Figure 8a. Use any resistors with values between 1000 Ω and 100 kΩ, and measure their exact values
before assembly.

Connect an ammeter across the output; its reading
represents the short-circuit current I_{Sc}, the maximum
current obtainable from the network. Note from the
Thevenin equivalent circuit of Figure 8b that I_{Sc} is
given by E_{Th}/R_{Th}.

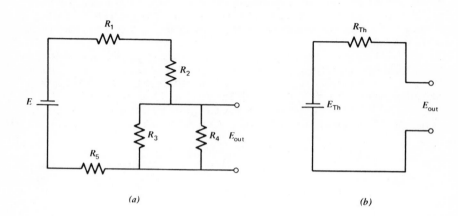

(a)   *(b)*

Figure 8.   Circuit (b) is the Thevenin equivalent of circuit (a).

With the help of an electronic voltmeter, measure the output voltage of the circuit. This "open-circuit" voltage is E_{Th}. From the values of I_{Sc} and E_{Th}, calculate R_{Th} (= Z_{Th}).

The procedure for calculating the Thevenin quantities from known component values is described in Chapter III. Make such a calculation for your circuit, and compare with the measured values.

Additional Experiments

Check any other two points in the circuit, and compare theoretical and experimental Thevenin quantities.

It is possible to measure R_{Th} by connecting a variable resistor at the output of the circuit and monitoring the voltage across it. When this voltage is equal to $E_{Th}/2$, the value of the resistance is equal to R_{Th}. Make this determination, and compare it with the value obtained by the short-circuit method. Which method seems to you most desirable, and why?

EXPERIMENT 6

Operational Amplifiers in Open-Loop

Objective: To examine the general behavior of open-loop operational amplifiers and their limitations.

References:

(1)   Chapter IV
(2)   J. G. Graeme, G. E. Tobey, and L. P. Huelsman (eds.) "Operational Amplifiers" (McGraw-Hill, N.Y., 1971)
(3)   Literature from manufacturers of operational amplifiers.

Theory: Operational amplifiers are high gain, differential amplifiers of negligible current input.   In this experiment, feedback will not be used, and hence the amplifiers will be, most of the time, at one of their voltage limits (saturated), a direct result of the extremely high gain.

Procedure

(1)   For this experiment, the common model 741 op amp can be used. A summary of some of its properties can be found in Table 4-2, but more details are available in the manufacturer's literature, which should be consulted. The pin diagrams for the three usual package types are given in Figure 9.

(2)   Construct the circuit shown in Figure 10a. Note that the upper resistor pair constitutes a voltage divider, generating a fixed voltage of about 1 V. (What should the resistor values be for exactly 1 V?) The lower resistor pair gives about 0 to 2 V, which can be varied by changing R_4.   The op amp in this circuit acts as a comparator and will go to either the + or - saturation output depending upon the relative values of E_1 and E_2.

(3)   Investigate the operation of the comparator by observing the output voltage for various input voltages E_2. Draw a graph to illustrate your results. Try to determine the resolution of the circuit by measuring the interval for E_2 within which the transition occurs.

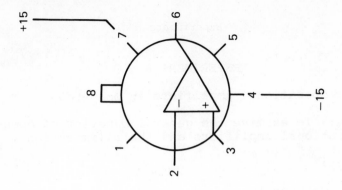

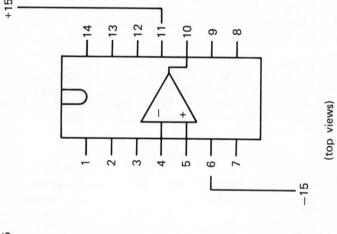

(top views)

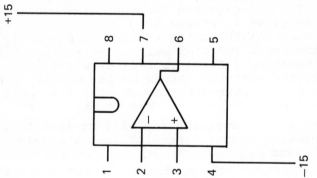

Figure 9.

362

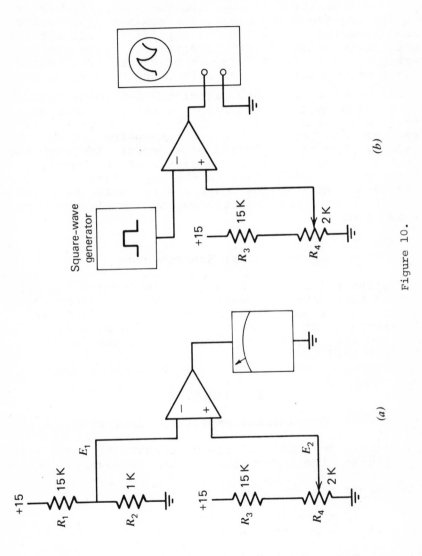

Figure 10.

363

(4)   Of great interest also are two additional
characteristics: (a) the slewing rate, which is the
speed of transition between the two saturated states,
and (b) the frequency limit, the maximum square-wave
input frequency for which the amplifier still attains
output saturation at both polarities. These can be observed by means of the circuit of Figure 10b. Using a
square-wave input of a few volts, and with E_2 at zero
volts, examine the scope patterns. Determine from them
the slewing rate, as the slope of the rise between -
and + saturation levels. Adjust both the oscillator
frequency and the oscilloscope ranges for convenient
readout. For the bandwidth measurement, increase the
input frequency until the amplitude of the output
starts to fall off. This is the upper limit of the
bandwidth. What is the lower limit? Note the shape of
the output waveforms for the various conditions mentioned above.

Additional Experiments

Using Figure 4-30 as a guide, determine the other
dynamic parameters indicated: rise time, overshoot,
and settling time. Find out from the manufacturer's
literature about specially optimized comparators. In
what way are they to be preferred over the 741 for
this service?

EXPERIMENT 7

Operational Amplifier Circuits

Objective: To study two basic circuits of operational
amplifiers: the integrator and the follower.

References:

(1)   Chapter IV
(2)   Graeme, Tobey, and Huelsman (loc. cit.)

Theory: The input-output relations for the two circuits
are:

(1)   Integrator:

$$E_{out} = - \frac{1}{RC} \int E_{in} dt$$

(2) Follower:

$$E_{out} = E_{in} \text{(exactly)}$$

Procedure

   (1) Construct the follower circuit of Figure 11a, using a 741 op amp. Measure the voltages at A and B, using an electronic voltmeter. The results should be equal within a few millivolts. Now measure the two voltages again, this time connecting a 1000-Ω load across the output terminals. Notice that the voltage at B is heavily affected, not so that at A. Explain.
   (2) Construct the integrator circuit of Figure 11b. Connect the 10-V recorder or meter between point A and ground, and measure the voltage at point B against ground with an electronic voltmeter. When S_2 is closed, the output at A should be zero. With S_1 closed and S_2 open, the circuit will integrate the input as a function of time. Calculate the theoretical slope of the output voltage versus time, based on the formula given above, and compare it with the observed rate of increase of voltage.
   Upon opening S_1, the circuit should retain its output voltage at a constant level. This is called the "hold" mode. The requirement for holding a voltage is the absence of drift. Measure the drift in your circuit by first integrating to approximately 1 V, then measuring the output change over a period of time when both S_1 and S_2 are open. Calculate the minimum voltage that can be held on the integrator without the drift affecting the precision by more than 1%/min. Also determine the longest time that one can hold a potential of 1 V before the error becomes larger than 1%.

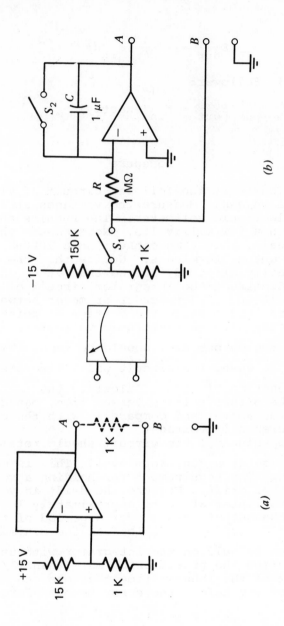

(a)

(b)

Figure 11.

Additional Experiments

(1)   Use the integrator as a means for measuring capacitance, and comment on its merits.

(2)   Feed into the integrator a low-frequency square wave (less than 1 Hz). Examine the output waveform and comment on the possible use of an integrator as a function generator.

(3)   Determine the response of the integrator to sine waves of 0.1, 1.0, and 10 Hz. Could the integrator be used as a filter? What happens to the DC component of the signal, if any?

EXPERIMENT 8

Power Supplies

Objective: To study a practical circuit employing transformer, diodes, and electrolytic capacitors.

Reference: Chapter VI.

Procedure

The circuit of Figure 12 is to be constructed one section at a time.

(1)   Assemble the power-line section, including the fuse, switch, and transformer. This should be approved by the instructor before connecting to the line. After approval, plug the circuit into the line, and measure the transformer output voltage with an oscilloscope or AC voltmeter. It should be a sine wave with a peak-to-peak voltage of $2 \times 25 \times \sqrt{2} = 71$ V, as observed on the scope, or 25 V, RMS, on the voltmeter, assuming that the rated voltage of the transformer is 25. If the observed voltage differs appreciably from this, it may be due to a nonstandard line voltage or to the tolerance of the transformer. Also check that the center tap (CT) does indeed divide the voltage into two equal parts.

(2)   Disconnect the transformer from the line, and add the two diodes and capacitor C_1. Measure the voltage across the capacitor with a DC voltmeter or with an oscilloscope. If a scope is used, you can observe the

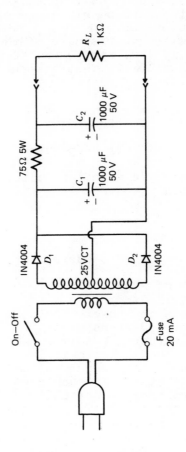

Figure 12.

residual AC (the "ripple"), by switching the scope to
the AC mode and increasing the sensitivity. Note the
wave-shape and frequency of the ripple, and comment upon your observations.

(3) Again disconnect the power line and complete
the circuit. Measure the output voltage and ripple
with and without the load, R_L, connected. The quality
of the power supply can be described by the magnitude
of the variation of the output on applying the load.
How good is your power supply? Does the ripple increase or decrease on connecting the load?

Additional Experiments

Determine the output impedance of the power supply
by the Thevenin method. Is it desirable for the output impedance to be high or low?

EXPERIMENT 9

Logic Gates

Objective: To study the properties of some logic gates
and to establish their truth tables.

Reference: Chapters VIII and IX.

Procedure

Figure 13 gives the pin diagrams of several common TTL gates. Note that pin 7 is ground and pin 14 is
V_{cc} (i.e., +5V) in all diagrams.

(1) Start with a 7400 NAND gate, using only one
of the four segments. Connect pins 7 and 14 appropriately. Connect a voltmeter between pin 3 and ground.
Ground both pins 1 and 2 and record the voltmeter reading. Now connect each of the input pins (1 and 2) separately and then together to the +5-V supply, and note
the results. Assemble your four observations into the
form of a truth table, comparable to those in Figures
8-4 and 8-5.

(2) Repeat (1) with each of the other basic types
of gates (7402, 7408, and 7432).

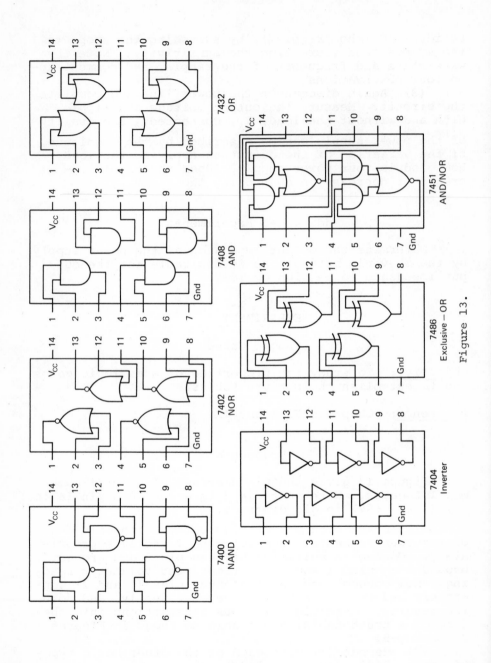

Figure 13.

(3)   Interconnect segments of 7408 and 7404 in the arrangement shown in Figure 8-8, and demonstrate that the resulting truth table is identical with that obtained from 7486.

(4)   The text shows several ways in which flip-flops can be synthesized from unit gates. Make up flip-flops of increasing sophistication following the diagrams of Figures 9-1, 9-2, and 9-4a, and prepare truth tables for each. Can you discern any advantages of one over the others?

(5)   The 7451 is an example of a more complex logic element on a single chip. (Nothing should be connected to pins 11 and 12 in this IC.) The two segments can be interconnected to form a particularly useful form of flip-flop known as a data latch. Look up the description of the 7475 latch, and reproduce its circuitry with the 7451 and establish its truth table.

Additional Experiments

An ingenious student can devise many complex logic systems using these gates, predict their truth tables, and then verify the same by experiment.

EXPERIMENT 10

Flip-flops and Counters

Objective: To illustrate RS-flip-flops and their use in counters and frequency division.

Reference: Chapter IX.

Procedure

Light-emitting diodes (LEDs) can be used to indicate the state of a logic element in place of the voltmeter specified in the previous experiment. To avoid burnout, a resistor must be connected in series with each LED;   470 Ω is a reasonable value.

(1)   Construct the circuit of Figure 14, using the manufacturer's literature to determine the pin connections. The two NAND gates form an RS flip-flop

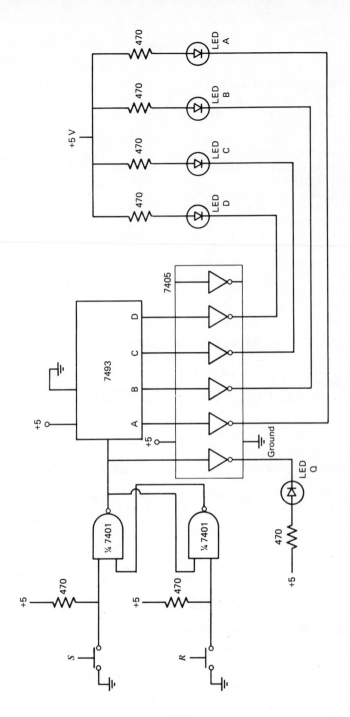

Figure 14.

which can be made to change state by pressing alter-
nately the R and S pushbutton switches. LED Q indicates a high state at Q. LEDs A, B, C, and D indicate
highs at consecutive outputs from the 7493 binary counter. The inverters (7405) serve simply as high-current
lamp drivers. (The 7404 inverters will not give
enough current for this; neither will the outputs of
the 7493.)

     (2) Test the operation of this circuit, and pre-
pare a table showing the overall system response, comparable to that in Figure 9-18.

     (3) Now connect a low-frequency square wave ($\sim$1
Hz) in place of the RS flip-flop, and note the sequence
of lamp signals. Can you suggest some uses for such a
system?

Additional Experiments

     Design a system using a monostable (use half of a
74121) instead of the RS flip-flop to respond to a single push-button instead of two.

Design a decimal counter along similar lines.

*   *   *

[From this point, the experiment directions will be
less detailed, requiring the student to hunt up the
specified references and modify the circuits there
presented, if necessary, to fit the equipment and components at hand.]

EXPERIMENT 11

Precision Full-wave Rectifier

Objective: To illustrate another op amp circuit.

Reference: J. G. Graeme: "Full-Wave Rectifier Needs
only Three Matched Resistors," Electronics, Aug. 8,
1974, p. 104.

Procedure

Construct the circuit and carefully adjust it for optimal operation. Such a circuit is sometimes called an "absolute value" circuit; why is this appropriate?

EXPERIMENT 12

Sine-Wave Oscillator

Objective: To construct a simple stable oscillator, and to illustrate the use of negative temperature co-efficient thermisters.

Reference: D. Hileman: "One op-amp Oscillator keeps Sine-wave Amplitude Constant," Electronics, June 24, 1976, p. 107.

Procedure

Construct the circuit and test it for amplitude stability as the load is changed, and for frequency stability versus time. A frequency counter can be used for the latter purpose. Load resistors of 1000 Ω or greater are appropriate.

EXPERIMENT 13

Active Filters

Objective: To study active filters.

References: (1) Chapter V and Figures 5-10 and 5-11. (2) G. O. Moberg: "Multiple Feedback Filter has Low Q and High Gain," Electronics, December 9, 1976, page 97.

Procedure

(1) Design and construct a band-pass filter following the schematic of Figure 5-10c, setting $f_1 = f_2$, at a level of several kilohertz. Try several combina-

tions of R and C, including some where $C_1 < C_2$, $C_1 = C_2$, and $C_1 > C_2$. For each combination, test the performance by means of a variable oscillator and electronic AC voltmeter or oscilloscope, and plot the corresponding Bode diagram.

(2) Repeat, using the circuitry of Figure 5-11c. Reference (2) will give you useful suggestions about this circuit (with one added resistor).

(3) Compare critically your results in this experiment with your prior results on passive filters in Experiment 4.

EXPERIMENT 14

Oscilloscope Multiplexor

Objective: To explore multiplexing, timing, and further use of the oscilloscope.

References: (1) Chapter IV. (2) M. L. Fichtenbaum: "Scope Display of Light Signals helps Debug Sequential Logic," Electronics, Dec., 25, 1975,p. 75. (3) C. S. Pepper: "Chopping Mode Improves Multiple Trace Display," Electronics, Oct. 14, 1976, p. 101.

Comments: In many experimental situations, it is very convenient, sometimes essential, to be able to observe several variables simultaneously on an oscilloscope screen. If the variables are in the analog domain, then a multiplexor can be designed around a programmable amplifier, such as that shown in Figure 4-19. Digital signals are more conveniently handled through digital logic devices. Reference (3) gives an excellent circuit for this purpose.

Procedure

Construct the Pepper circuit, adapted if necessary to match the oscilloscope at hand. As an additional experiment, design and construct an analog multiplexor using the Harris HA-2400 amplifier.

EXPERIMENT 15

A Protective Device

Objective: To study a protective device interfaced
with the power line.

Reference: R. J. Patel: "555 Timer Isolates Equipment
from Excessive Line Voltage," Electronics, Sept. 15,
1977, p. 116.

Procedure

After constructing the circuit, test it using a
variable transformer to alter the line voltage. For
the voltage regulator in the diagram, use a 7812 IC.

EXPERIMENT 16

Auto Anti-Theft Device

Objective: To illustrate the use of electronics in
everyday life.

Reference: G. L. Grundy: "Engine Staller Thwarts Car
Thieves," Electronics, Dec. 23, 1976, p. 71.

Procedure

Construct the circuit as indicated.

EXPERIMENT 17

Triangular Wave Generator

Objective: To illustrate the use of timers and associated circuits.

Reference: D. M. Gualtieri: "Triangular Waves from
555 have Adjustable Symmetry," Electronics, Jan. 8,
1976, p. 111.

Procedure

Construct the circuit indicated and test its behavior.

EXPERIMENT 18

Noise Generator

Objective: To construct a source of white noise, using shift registers and exclusive-OR gates.

References: (1) J. Maxwell: "MOS Op Amps form "Pink Noise" Source," Electronics, March 31, 1977, p. 118. (2) M. Damashek: "Shift Register with Feedback Generates White Noise," Electronics, May 27, 1976, p. 107.

Comment: A generator that will produce random impulses over a wide frequency range is useful in evaluating audio equipment and filters. The references give circuits for constructing such a noise generator. The output is called "pseudo-random" because it will eventually repeat the same sequence of pulses, but with a period much longer than most experiments.

Procedure

Construct the noise generator described in one of the references. Test its utility in checking the action of both passive and active filters of various types.

EXPERIMENT 19

Analog-Digital Conversion

Objective: To explore some relations between analog and digital signals.

References: (1) Chapter XI. (2) D. H. Sheingold (ed.): "Analog - Digital Conversion Handbook, " Analog Devices, 1972, p. I-68, ff. (3) R. L. Morris and J. R. Miller (eds.): "Designing with TTL Inte-

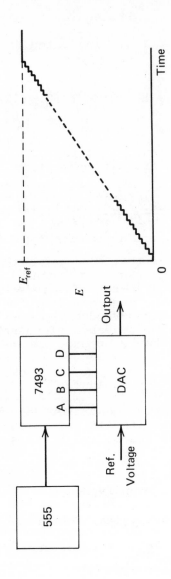

Figure 15.

grated Circuits," McGraw-Hill, N.Y., 1971, p. 161, ff,
and p. 267, ff.   (4)   E. R. Hnatek: "A User's Handbook of Semiconductor Memories," Wiley, N.Y., 1977,
p. 294.

Comments: A digital-to-analog converter (DAC) can be
used to advantage to generate a "staircase" function,
i.e., one in which the voltage rises in a series of
discrete steps, usually of equal height and equal duration.   For this purpose, it is driven by successive
digital outputs from a binary counter, as in Figure 15,
where a 555 timer, connected as a free-running square-
wave source, feeds a 7493 binary counter.   The 1408 is
an 8-bit multiplying DAC, which means that the output
is equal to a reference voltage multiplied by the binary input number.   The staircase will start at zero
volts at zero time, and terminate at E_{Ref} at a time
determined by the frequency of the 555.   The interval
between is divided into $2^8 = 256$ equal steps, which
may be so small that the result is indistinguishable
on the oscilloscope from a diagonal straight line.   If
an up-down counter such as the 74193 is substituted
for the 7493 (Ref. 3), a simple logic step can result
in a stepped triangular wave (Ref. 2).   Furthermore,
suitable manipulation between counter and DAC can
transform the triangular wave into any other repetitive function.   The MM522 is a sine-function generator
(a special-purpose ROM, Ref. 4) that can be used as in
Figure 16 to produce a highly precise sine wave by this
technique.

Procedure

The 8-bit DAC for this experiment can be the
MC1408L8 (Motorola), AD7523 (Analog Devices), DAC-08
(Datel), or equivalent.   In place of the 555, an external square-wave or pulse generator can be used if
desired.

(1) First assemble the ramp generator of Figure
15, and observe its output on the oscilloscope.   By
adjusting the scope controls and the frequency of the
555, you should be able to see the ramp either as a
series of steps or as an apparently continuous straight
line.   This can also be plotted on a strip-chart or

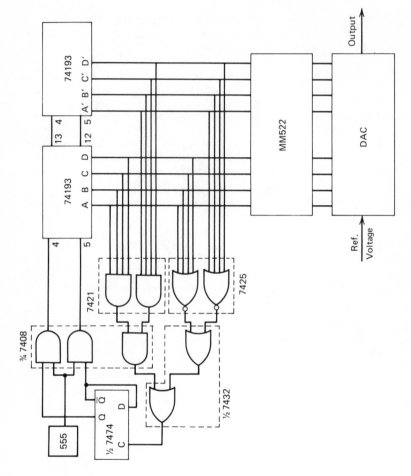

Figure 16.

X-Y recorder. If any nonlinearity appears, do you consider it the fault of your generator or of the recorder?

(2) Connect up the sine-wave generator of Figure
16, and make similar observations. What limits the
frequency, at both high and low ends, of the sine wave
that can be generated?

APPENDIX II

TABLE OF COMMON LOGARITHMS

N	0	1	2	3	4	5	6	7	8	9
10	0000	0043	0086	0128	0170	0212	0253	0294	0334	0374
11	0414	0453	0492	0531	0569	0607	0645	0682	0719	0755
12	0792	0828	0864	0899	0934	0969	1004	1038	1072	1106
13	1139	1173	1206	1239	1271	1303	1335	1367	1399	1430
14	1461	1492	1523	1553	1584	1614	1644	1673	1703	1732
15	1761	1790	1818	1847	1875	1903	1931	1959	1987	2014
16	2041	2068	2095	2122	2148	2175	2201	2227	2253	2279
17	2304	2330	2355	2380	2405	2430	2455	2480	2504	2529
18	2553	2577	2601	2625	2648	2672	2695	2718	2742	2765
19	2788	2810	2833	2856	2878	2900	2923	2945	2967	2989
20	3010	3032	3054	3075	3096	3118	3139	3160	3181	3201
21	3222	3243	3263	3284	3304	3324	3345	3365	3385	3404
22	3424	3444	3464	3483	3502	3522	3541	3560	3579	3598
23	3617	3636	3655	3674	3692	3711	3729	3747	3766	3784
24	3802	3820	3838	3856	3874	3892	3909	3927	3945	3962
25	3979	3997	4014	4031	4048	4065	4082	4099	4116	4133
26	4150	4166	4183	4200	4216	4232	4249	4265	4281	4298
27	4314	4330	4346	4362	4378	4393	4409	4425	4440	4456
28	4472	4487	4502	4518	4533	4548	4564	4579	4594	4609
29	4624	4639	4654	4669	4683	4698	4713	4728	4742	4757
30	4771	4786	4800	4814	4829	4843	4857	4871	4886	4900

N	0	1	2	3	4	5	6	7	8	9
31	4914	4928	4942	4955	4969	4983	4997	5011	5024	5038
32	5051	5065	5079	5092	5105	5119	5132	5145	5159	5172
33	5185	5198	5211	5224	5237	5250	5263	5276	5289	5302
34	5315	5328	5340	5353	5366	5378	5391	5403	5416	5428
35	5441	5453	5465	5478	5490	5502	5514	5527	5539	5551
36	5563	5575	5587	5599	5611	5623	5635	5647	5658	5670
37	5682	5694	5705	5717	5729	5740	5752	5763	5775	5786
38	5798	5809	5821	5832	5843	5855	5866	5877	5888	5899
39	5911	5922	5933	5944	5955	5966	5977	5988	5999	6010
40	6021	6031	6042	6053	6064	6075	6085	6096	6107	6117
41	6128	6138	6149	6160	6170	6180	6191	6201	6212	6222
42	6232	6243	6253	6263	6274	6284	6294	6304	6314	6325
43	6335	6345	6355	6365	6375	6385	6395	6405	6415	6425
44	6435	6444	6454	6464	6474	6484	6493	6503	6513	6522
45	6532	6542	6551	6561	6571	6580	6590	6599	6609	6618
46	6628	6637	6646	6656	6665	6675	6684	6693	6702	6712
47	6721	6730	6739	6749	6758	6767	6776	6785	6794	6803
48	6812	6821	6830	6839	6848	6857	6866	6875	6884	6893
49	6902	6911	6920	6928	6937	6946	6955	6964	6972	6981
50	6990	6998	7007	7016	7024	7033	7042	7050	7059	7067
51	7076	7084	7093	7101	7110	7118	7126	7135	7143	7152
52	7160	7168	7177	7185	7193	7202	7210	7218	7226	7235
53	7243	7251	7259	7267	7275	7284	7292	7300	7308	7316
54	7324	7332	7340	7348	7356	7364	7372	7380	7388	7396
55	7404	7412	7419	7427	7435	7443	7451	7459	7466	7474
56	7482	7490	7497	7505	7513	7520	7528	7536	7543	7551
57	7559	7566	7574	7582	7589	7597	7604	7612	7619	7627
58	7634	7642	7649	7657	7664	7672	7679	7686	7694	7701
59	7709	7716	7723	7731	7738	7745	7752	7760	7767	7774
60	7782	7789	7796	7803	7810	7818	7825	7832	7839	7846
61	7853	7860	7868	7875	7882	7889	7896	7903	7910	7917
62	7924	7931	7938	7945	7952	7959	7966	7973	7980	7987
63	7993	8000	8007	8014	8021	8028	8035	8041	8048	8055
64	8062	8069	8075	8082	8089	8096	8102	8109	8116	8122
65	8129	8136	8142	8149	8156	8162	8169	8176	8182	8189
66	8195	8202	8209	8215	8222	8228	8235	8241	8248	8254
67	8261	8267	8274	8280	8287	8293	8299	8306	8312	8319
68	8325	8331	8338	8344	8351	8357	8363	8370	8376	8382
69	8388	8395	8401	8407	8414	8420	8426	8432	8439	8445

N	0	1	2	3	4	5	6	7	8	9
70	8451	8457	8463	8470	8476	8482	8488	8494	8500	8506
71	8513	8519	8525	8531	8537	8543	8549	8555	8561	8567
72	8573	8579	8585	8591	8597	8603	8609	8615	8621	8627
73	8633	8639	8645	8651	8657	8663	8669	8675	8681	8686
74	8692	8698	8704	8710	8716	8722	8727	8733	8739	8745
75	8751	8756	8762	8768	8774	8779	8785	8791	8797	8802
76	8808	8814	8820	8825	8831	8837	8842	8848	8854	8859
77	8865	8871	8876	8882	8887	8893	8899	8904	8910	8915
78	8921	8927	8932	8938	8943	8949	8954	8960	8965	8971
79	8976	8982	8987	8993	8998	9004	9009	9015	9020	9025
80	9031	9036	9042	9047	9053	9058	9063	9069	9074	9079
81	9085	9090	9096	9101	9106	9112	9117	9122	9128	9133
82	9138	9143	9149	9154	9159	9165	9170	9175	9180	9186
83	9191	9196	9201	9206	9212	9217	9222	9227	9232	9238
84	9243	9248	9253	9258	9263	9269	9274	9279	9284	9289
85	9294	9299	9304	9309	9315	9320	9325	9330	9335	9340
86	9345	9350	9355	9360	9365	9370	9375	9380	9385	9390
87	9395	9400	9405	9410	9415	9420	9425	9430	9435	9440
88	9445	9450	9455	9460	9465	9469	9474	9479	9484	9489
89	9494	9499	9504	9509	9513	9518	9523	9528	9533	9538
90	9542	9547	9552	9557	9562	9566	9571	9576	9581	9586
91	9590	9595	9600	9605	9609	9614	9619	9624	9628	9633
92	9638	9643	9647	9652	9657	9661	9666	9671	9675	9680
93	9685	9689	9694	9699	9703	9708	9713	9717	9722	9727
94	9731	9736	9741	9745	9750	9754	9759	9763	9768	9773
95	9777	9782	9786	9791	9795	9800	9805	9809	9814	9818
96	9823	9827	9832	9836	9841	9845	9850	9854	9859	9863
97	9868	9872	9877	9881	9886	9890	9894	9899	9903	9908
98	9912	9917	9921	9926	9930	9934	9939	9943	9948	9952
99	9956	9961	9965	9969	9974	9978	9983	9987	9991	9996

APPENDIX III

DECIBEL TABLE

dB	Voltage Gain	Power Gain	Voltage Loss	Power Loss
0.0	1.00	1.00	1.00	1.00
0.1	1.01	1.02	0.99	0.98
0.2	1.02	1.05	0.98	0.96
0.3	1.04	1.07	0.97	0.93
0.4	1.05	1.10	0.96	0.91
0.5	1.06	1.12	0.94	0.89
0.6	1.07	1.15	0.93	0.87
0.7	1.08	1.17	0.92	0.85
0.8	1.10	1.20	0.91	0.83
0.9	1.11	1.23	0.90	0.81
1.0	1.12	1.26	0.89	0.79
2.0	1.26	1.58	0.79	0.63
3.0	1.41	2.00	0.71	0.501
4.0	1.58	2.51	0.63	0.398
5.0	1.78	3.16	0.56	0.316
6.0	2.00	3.98	0.501	0.251

dB	Voltage Gain	Power Gain	Voltage Loss	Power Loss
7.0	2.24	5.01	0.447	0.200
8.0	2.51	6.31	0.398	0.158
9.0	2.82	7.94	0.355	0.126
10.0	3.16	10.00	0.316	0.100
20.0	10.00	100.00	0.100	0.010

Note that the voltage ratio for 3 dB is $\sqrt{2}$, while 6 dB corresponds to a ratio of 2. For power, the ratios are 2 and 4, respectively. If a value not in the table is required, one should break down the decibel number into a sum of known values, take them from the table, and multiply the results. Thus 43.5 dB as a voltage gain can be written as 20 + 20 + 3 + 0.5 corresponding to gains of 10, 10, 1.41, and 1.06. The net gain is given by $10 \times 10 \times 1.41 \times 1.06 = 149.5$.

APPENDIX IV

TRIGONOMETRIC TABLES

Angle		Sine	Cosine	Tangent
Degrees	Radians			
0	0	0	1	0
30	$\pi/6$	1/2	$\sqrt{3}/2$	$1/\sqrt{3}$
45	$\pi/4$	$\sqrt{2}/2$	$\sqrt{2}/2$	1
60	$\pi/3$	$\sqrt{3}/2$	1/2	$\sqrt{3}$
90	$\pi/2$	1	0	∞
120	$2\pi/3$	$\sqrt{3}/2$	$-1/2$	$-\sqrt{3}$
135	$3\pi/4$	$\sqrt{2}/2$	$-\sqrt{2}/2$	-1
150	$5\pi/6$	1/2	$-\sqrt{3}/2$	$-1/\sqrt{3}$
180	π	0	-1	0
$-a$	$--$	$-\sin a$	$\cos a$	$-\tan a$
$90 + a$	$--$	$\cos a$	$-\sin a$	$-\cotan a$
$90 - a$	$--$	$\cos a$	$\sin a$	$\cotan a$

APPENDIX V

CAPACITIVE IMPEDANCE

The table gives the impedance (capacitive reactance) of selected capacitances, calculated from the
formula $Z_c = X_c = 1/(2\pi f C)$. Values less than 0.1 Ω
are not likely to be useful, and so are not given. Actual capacitors will obey this relation only over limited frequency ranges, because of inductive and dis-
sipative effects. The figure* shows the approximate
ranges over which various types are useful.

*Reproduced, with permission, from H. W. Ott, *Noise Reduction Tech-
niques in Electronic Systems,* (Wiley, New York, 1976; copyright:
Bell Telephone Laboratories, Inc.), p. 117.

CAPACITIVE IMPEDANCE VALUES[†]

Capacitance (μF)	50 Hz	100 Hz	1 kHz	10 kHz	100 kHz	1 MHz	10 MHz
0.001	3.2 MΩ	1.6 MΩ	160. kΩ	16. kΩ	1.6 kΩ	160. Ω	16. Ω
0.005	640. kΩ	320. kΩ	32. kΩ	3.2 kΩ	320. Ω	32. Ω	3.2 Ω
0.01	320. kΩ	160. kΩ	16. kΩ	1.6 kΩ	160. Ω	16. Ω	1.6 Ω
0.05	64. kΩ	32. kΩ	3.2 kΩ	320. Ω	32. Ω	3.2 Ω	0.32 Ω
0.1	32. kΩ	16. kΩ	1.6 kΩ	160. Ω	16. Ω	1.6 Ω	0.16 Ω
0.5	6.4 kΩ	3.2 kΩ	320. Ω	32. Ω	3.2 Ω	0.32 Ω	--
1.0	3.2 kΩ	1.6 kΩ	160. Ω	16. Ω	1.6 Ω	0.16 Ω	--
5.0	640. Ω	320. Ω	32. Ω	3.2 Ω	0.32 Ω	--	--
10.0	320. Ω	160. Ω	16. Ω	1.6 Ω	0.16 Ω	--	--
50.0	64. Ω	32. Ω	3.2 Ω	0.32 Ω	--	--	--
100.0	32. Ω	16. Ω	1.6 Ω	0.16 Ω	--	--	--
500.0	6.4 Ω	3.2 Ω	0.32 Ω	--	--	--	--

[†]Values under 0.1 Ω are not indicated.

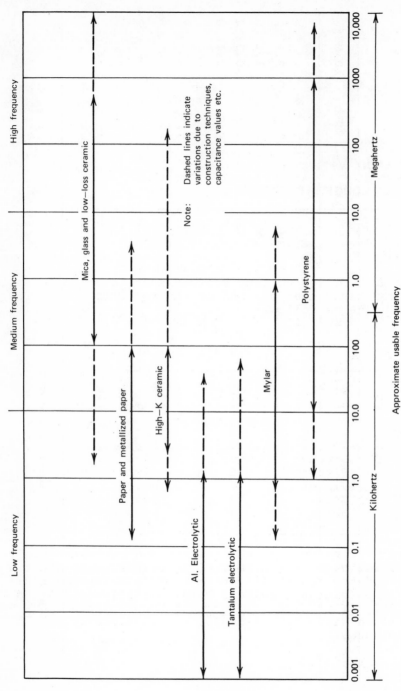

COLOR CODE FOR RESISTORS

The values of low-wattage resistors are coded by a series of colored bands near one end of the resistor body. The color code is as follows:

Color	As Digit	As Number of Zeros
Black	0	None
Brown	1	One
Red	2	Two
Orange	3	Three
Yellow	4	Four
Green	5	Five
Blue	6	Six
Violet	7	Seven
Grey	8	Eight
White	9	Nine

The first two bands from the nearest end represent digits, the next indicates the number of zeros following the two digits, and the fourth (if present) is a tolerance indicator. The tolerance code is gold, 5%; silver, 10%; no color, 20%. Thus a resistor banded brown-black-green-silver has a value of 1-0-00000, or 1,000,000 Ω, within 10% tolerance.

APPENDIX VII

LAPLACE TRANSFORMS

The Laplace transform of a function $F(t)$ is defined by the integral

$$\overline{F}\, s \;=\; \int_0^\infty F(t)\, e^{-st}\, dt$$

The inverse transform is given by

$$F(t) \;=\; \frac{1}{2\pi j} \int_{c-j\infty}^{c+j\infty} \overline{F}(s)\, e^{st}\, ds$$

The table gives pairs of equivalent expressions for selected functions. Extensive tables are available in many reference works.

LAPLACE TRANSFORMS

$\overline{F}(s)$	$F(t)$
$1/s$	Unit step
$1/s^2$	t
$1/s^n$	$t^{n-1}/(n-1)!$
$1/\sqrt{s}$	$1/\sqrt{\pi t}$
$1/(s + a)$	$\exp(-at)$
$\dfrac{1}{s(s + a)}$	$(1/a)[1 - \exp(-at)]$
$\dfrac{1}{s(s + a)(s + b)}$	$\dfrac{1}{ab}\left[1 + \dfrac{1}{a - b}\,[b\exp(-at) - a\exp(-bt)]\right]$
$\dfrac{1}{(s + a)(s + b)(s + c)}$	$\dfrac{\exp(-at)}{(b - a)(c - a)} + \dfrac{\exp(-bt)}{(a - b)(c - b)} + \dfrac{\exp(-ct)}{(a - c)(b - c)}$
$\dfrac{1}{(s + a)^2}$	$t\exp(-at)$

$$\frac{1}{(s+a)^n} \qquad \frac{t^{n-1}\exp(-at)}{(n-1)!}$$

$$\frac{1}{s(s+a)^2} \qquad \frac{1}{a^2}[1-\exp(-at)-at\exp(-at)]$$

$$\frac{1}{s^2+a^2} \qquad \frac{1}{a}\sin at$$

$$\frac{1}{s(s^2+a^2)} \qquad \frac{1}{a^2}(1-\cos at)$$

$$\frac{s}{s^2+a^2} \qquad \cos at$$

$$\frac{1}{(s^2+a^2)^2} \qquad \frac{1}{2a^3}(\sin at - at\cos at)$$

$$\frac{1}{(s+a)^2+b^2} \qquad \frac{1}{b}\exp(-at)\sin bt$$

$$\frac{1}{s^2-a^2} \qquad \frac{1}{a}\sinh at$$

$$\frac{s}{s^2-a^2} \qquad \cosh at$$

$\bar{F}(s)$	$F(t)$
$\dfrac{s}{(s+a)^2}$	$(1 - at)\exp(-at)$
$\dfrac{s}{(s+a)(s+b)}$	$\dfrac{1}{a-b}[a\exp(-at) - b\exp(-bt)]$
$\dfrac{s}{(s^2+a^2)^2}$	$\dfrac{1}{2a}\,t\sin at$
$\dfrac{s+a}{s(s+b)}$	$\dfrac{a}{b} - \dfrac{a-b}{b}\exp(-bt)$
$\dfrac{s+a}{(s+b)^2}$	$[1 + (a - b)t]\exp(-bt)$
$\dfrac{s^2 - a^2}{(s^2+a^2)^2}$	$t\cos at$

APPENDIX VIII

SUGGESTIONS FOR FURTHER READING

Brophy, J. J.: *Basic Electronics for Scientists.* 3rd
   ed. McGraw-Hill, New York, 1977.
Connelly, J. A. (Ed.): *Analog Integrated Circuits.*
   Wiley, New York, 1975.
Diefenderfer, A. J.: *Principles of Electronics Instru-*
   mentation. Saunders, Philadelphia, 1972.
_____, *Basic Techniques in Electronic Instrumentation.*
   [A laboratory text.] Saunders, Philadelphia, 1972.
Graeme, J. G., Tobey, G. E., and Huelsman, L. P. (Eds.):
   Operational Amplifiers, Design and Applications.
   McGraw-Hill, New York, 1971.
Graeme, J. G.: *Applications of Operational Amplifiers:*
   Third Generation Techniques. McGraw-Hill, New
   York, 1973.
Greenfield, J. D.: *Practical Digital Design Using ICs.*
   Wiley, New York, 1977.
Hnatek, E. R.: *Applications of Linear Integrated Cir-*
   cuits. Wiley, New York, 1975.
Morris, R. L., and Miller, J. R. (Eds.): *Designing*
   with TTL Integrated Circuits. McGraw-Hill, New
   York, 1971.
Ott, H. W.: *Noise Reduction Techniques in Electronic*
   Systems. Wiley, New York, 1976.
Shacklette, L. W., and Ashworth, H. A.: *Using Digital*
   and Analog Integrated Circuits. [A laboratory
   text.] Prentice-Hall, Englewood Cliffs, NJ, 1978.
Smith, J. I.: *Modern Operational Circuit Design.*
   Wiley, New York, 1971.

In addition to the above books, we highly recommend the catalogs and applications manuals available from many manufacturers of integrated circuits. Those that have been particularly helpful to the authors are published by the following firms:

Analog Devices
P.O. Box 280, Norwood, MA 02062

Burr-Brown Research Corp.
International Airport Industrial Park, Tucson, AZ 85734

Fairchild Semiconductor
464 Ellis St., Mountain View, CA 94042

Harris Semiconductor
P.O. Box 883, Melbourne, FL 32901

Motorola Semiconductor Products
Box 20912, Phoenix, AZ 85036

National Semiconductor Corp.
2900 Semiconductor Drive, Santa Clara, CA 95051

Precision Monolithics, Inc.
1500 Space Park Dr., Santa Clara, CA 95050

Signetics Corporation
811 East Arques Ave., Sunnyvale, CA 94086

Siliconix, Inc.
2201 Laurelwood Rd., Santa Clara, CA 95054

INDEX